Contributions to Economics

The series *Contributions to Economics* provides an outlet for innovative research in all areas of economics. Books published in the series are primarily monographs and multiple author works that present new research results on a clearly defined topic, but contributed volumes and conference proceedings are also considered. All books are published in print and ebook and disseminated and promoted globally. The series and the volumes published in it are indexed by Scopus and ISI (selected volumes).

More information about this series at https://link.springer.com/bookseries/1262

Zhiyi Liu

Principles of Digital Economics

Innovation Theory in the Age of Intelligence

Springer

Zhiyi Liu
SenseTime Group Limited
Shanghai, China

ISSN 1431-1933 ISSN 2197-7178 (electronic)
Contributions to Economics
ISBN 978-981-16-9019-8 ISBN 978-981-16-9020-4 (eBook)
https://doi.org/10.1007/978-981-16-9020-4

This Springer imprint is published by the registered company Springer Nature Singapore Pte Ltd.
The registered company address is: 152 Beach Road, #21-01/04 Gateway East, Singapore 189721, Singapore

Preface

From Global Financial Capitalism to Digital Economic Enlightenment

For the world economy, 2018 has historical significance. We are at a critical point in history, both in terms of the development of the global economy and in terms of the development of the Chinese economy.

On the one hand, the global economy is experiencing low economic growth due to weak income growth and income distribution problems, the effects of which have spread from the economic to the political and social spheres. The global financial crisis since 2008 has not only not ended completely but has also led to an evolution of the internal logic of the world political situation and has triggered a continued escalation of trade between the USA and China and a conservatization of Western politics. The Trump administration announced a second round of tariff sanctions against Chinese imports on 17 September 2018. $200 billion of Chinese imports were covered in this round of tariff sanctions. In other words, the global economy has not recovered from the end of the phase of the financial crisis a decade ago. On the contrary, the rising tide of counter-globalization and the prolongation of trade wars are being promoted. On the other hand, China is entering its fourth decade of reform and opening up, an important period in its history. At the same time, the economic situation at home and abroad has undergone significant changes. In the last decade, China has gone from being the world's third-largest exporter to the world's number one exporter. At the same time, China is changing the way it has promoted economic development over the past decade by increasing leverage significantly. The pursuit of high growth is being transformed into high-quality development. We can see the implementation of government policies to promote deleveraging to prevent systemic financial risks, precise poverty alleviation, and improvement of the ecological environment. These are all based on the improvement of the quality of the overall economic development. Within the past decade, the model of relying on exports and real estate to drive the economy was changed. The internal logic of economic growth has been realized through the removal of production capacity and structural transformation. In short, the logic of GDP-driven economic development is receding from the stage of history. The Chinese economy is undergoing a transition from high growth to high-quality development.

Against this backdrop, an important trend is that the development of the new economy, especially the digital economy, is becoming an increasingly important economic phenomenon. The transition from an industrial economy to a digital economy is accelerating and the whole of human civilization is shifting toward a digital world. The digital economy has become a new form of economic and social development after the agricultural and industrial economies. It is not only the main driving force behind the economic transformation of individual countries but has also become the high point of a new round of industrial competition. From a global perspective, China, as the second-largest digital economy in the world, has seen its digital economy grow at a compound annual growth rate of 25.1%. It is roughly three times the growth rate of the digital economy in the USA, Japan, and the UK. At the same time, it is also three times the compound growth rate of the real economy of the USA and the UK, for example. As technologies such as artificial intelligence, cloud computing, blockchain, and the Internet of Things develop, we can see that the digital economy is not only driving the construction of new factors of production and infrastructure but also the shift of human society as a whole toward a smart age. In fact, we have seen China's digital economy develop over the past two decades with the Internet at its core. And when technologies such as blockchain are developed, the Internet of information is being transformed into the Internet of value. In addition, we are seeing a change in the internal logic of the digital economy, from a focus on "innovation in business models" to a focus on "change in production relations." This is the next important change in the development of the digital economy.

In short, the digital economy is becoming an important area distinct from the real and financial (virtual) economy. It also plays an important role in empowering the real economy as well as changing the traditional financial landscape through fintech. After writing my book "Origins: From the Turing Test to the Blockchain Consensus," as a digital economist and technology thinker, I also shifted my focus from technology thinking to thinking about the impact of the digital economy on society and economic development. At the same time, all relevant results are studied in the framework of the original subdiscipline of economics "Digital Economics," so that a new paradigm of thinking is offered to understand the phenomena in the digital economy. On 17 May 2016, General Secretary Xi Jinping stated at a symposium on social science work "This is an era that needs theories and can certainly produce them. This is an era that needs ideas and can certainly produce them." I think we, as Chinese economists, should also have such a sense of mission for the times. We should uphold the spirit of seeking truth from facts to explore economic theories that are in line with the development of China's national conditions and with an international perspective.

In what follows, I will explore the background and logic of thinking about the discipline of digital economics from the following three perspectives. At the same time, I hope that the reader will be able to gain a first insight into the content framework and ideas of the book through the introduction of this preface.

Global Financial Capitalism

In 2008, the financial crisis broke out, the first truly global financial crisis since the Second World War. Alan Greenspan, chairman of the Federal Reserve, said that it was "the financial crisis of the century" and even "the most severe financial crisis in human history." He also referred to "the near instantaneous collapse of short-term financial markets around the world during the 2008 financial tsunami, although a much larger collapse of global economic activity occurred in the 1930s". This was an unprecedented and alarming phenomenon. The 2008 financial crisis revealed the harsh truth to all those who were overly optimistic about the development of modern capitalism that the structural problems of the world economy were difficult to root out and that all existing macroeconomic policies had failed. When the economy fell sharply and unemployment rose, the financial crisis brought about not only a contraction of the real economy but also a reckless quantitative easing of monetary policy by the central banks (although it had little effect). In other words, the global financial crisis made everyone start to realize that there was something wrong with the existing economic institutional arrangements. At the same time, everyone was virtually helpless to do anything about the problem.

The current problems of the global economy are in fact not only the outbreak of a crisis brewing in global finance capitalism but also the manifestation of a new form of the global economic system. On the one hand, the development of globalization since the 1970s has led to a shift from a country-specific economy to a global economy, which has led to the gradual formation of an economic ecology dominated by transnational corporations. Such an ecology can lead to the following scenarios. When an economic crisis breaks out, it spreads rapidly through economic networks to entities in every country. Ultimately, an amplification of the financial crisis will occur. On the other hand, the capitalist system has entered the era of global capitalism, which refers to an era in which global money and credit creation are central. In this era, the real economy is replaced by the money and financial markets, which become the mainstay of the economy. In an international monetary system dominated by the US dollar standard, global monetary and financial markets have gradually developed a highly unified market mechanism. Both the US dollar and the financial policy represented by the Federal Reserve influence the money supply, the level of inflation, international trade, and the monetary policy of national economies. As a result, a globalized financial capitalism has emerged.

The "derealization" of the global economy is the biggest change brought about by global financial capitalism, which has three main aspects.

First, the financial markets, represented by the money, bond, and stock markets, have become the most important markets dominating the global economic system. Capitalism has shifted from commodity capitalism and industrial capitalism to financial capitalism, which has brought about innovative financial instruments such as venture capital mechanisms, equity incentives, and asset securitization. At the same time, it has brought about the development of information technology and a new era of global digitalization. In particular, money, as a special commodity of general equivalence, has become a symbol of credit in the form of "paper money" or even "digital money." Since the collapse of the Bretton Woods system, the expansion of monetary credit, represented by the US dollar, has brought about a transformation of the global economy into financial capitalism, with the financial markets evolving into a complex and diverse market that influences the global industrial division of labor and the trading system.

Second, three key price signals determine the allocation of resources, the structure of the economy, and the distribution of income in each country. These three price signals are interest rates, exchange rates, and asset prices, and they are also the fundamental variables of global financial capitalism. The three price signals are completely divorced not only from the fundamentals of the real economy but also from the macro policy implications of sovereign states. This has led not only to a lack of monetary and fiscal policy by governments and central banks but also to structural instability and uncertainty in the global economic system. With the development of financial engineering and financial innovation, the market for the issuance and trading of financial derivatives became a trading market in the global economy that exceeded the size of the real economy by tens of times. This led to the financial crisis of 2008 and the ensuing global trend toward trade conservatism.

Third, the ability to innovate has become the most important capability of a company, and creating value for shareholders through innovation has become the main task of companies in the era of global financial capitalism. The development of global financial capitalism has not only driven the birth of innovative companies, represented by the innovative companies in Silicon Valley, but also the development of an innovative culture based on P/E ratios, shareholder value, and corporate market capitalization. As a result, the global industrial model and industrial division of labor system have been reshaped based on such incentives. Furthermore, it has become the basis for a realistic system for the development of a global digital economy.

The above qualities make global financial capitalism a complex system of dynamic evolution, nonequilibrium, and nonlinearity. It is difficult for mainstream economics, which is based on the study of closed economic systems, to reveal the laws of operation and internal contradictions of global financial capitalism. Global financial capitalism is a truly open economic system, and any national or regional economy is forced to be involved in this open system. The existing theories are all based on theories founded under the closed economy model. Therefore, existing theories will partially or completely fail in the conditions of a global open economy. On the one hand, new economic phenomena deprive the core theories of modern economics of their explanatory power. For example, equilibrium price theory is relatively weak in explaining financial asset prices, and traditional market theory lacks insight into the markets for investment factors of production. On the other hand, the need for new theoretical frameworks in economics and the exploration of related theories continues to arise, especially in the area of research on the digital economy where new economic phenomena are needed. For example, the reallocation of economic factors brought about by blockchain and artificial intelligence, the reorganization of capital theory, and the limited nature of the traditional theoretical framework of the firm need to be addressed.

The new economic phenomenon required scholars to start rethinking economics based on such a new economic ecology and to think about the solutions for the human economy under global finance capitalism in order to establish a theoretical framework and paradigm different from that of traditional economics. This is the important context in which digital economics was born. What is particularly important to be mentioned is that the logic behind the emergence of global finance capitalism is the progressivist spirit of the West since the Enlightenment. The whole idea of mainstream Western economics was laid down by the Western philosophy of theory-only and empiricism. Contemporary mainstream Western economics is essentially along the tradition of the scientistic orientation of Cyril Muller Keynes. In other words, modern economics has been laid down by Western philosophical thought. At a time when the whole of human civilization is being driven by Western progressivism and instrumentalism toward the tidal wave of global finance capitalism, should we return to Eastern philosophical thought and humanistic concepts to rethink the basic philosophical concepts of the digital economy?

A New Era of Digital Economy

Global financial capitalism has not only led to a "derealization" of the global economy but has also laid the institutional basis for the development of the digital economy and the ground for innovation. The creation of mechanisms such as venture capital, the Growth Enterprise Market (represented by NASDAQ), and equity incentives by global financial capital has made the digital economy not only the high point of a new round of global industrial competition but also a new way of thinking about global economic development. Economist Zhou Qiren once said, "Today's digital and digitization allows direct access to consumption and exchange. It is more important the direct access to production and its profound transformation of all economic and social activities." The digital economy does not just mean the digitization of industry or the industrialization of digital but represents an abstract digital culture and the inherent development logic and structural changes that digital technology brings to the entire human economy. The key to the shift from a resource-driven to a knowledge-driven and technology-driven economy for China is the emphasis on the digital economy. Specifically, the new economic ecology shaped by the digital economy is characterized by the following three aspects:

First, digitized knowledge and information are becoming the most central factors of production. When artificial intelligence develops, we can expect to see a massive transformation of the existing division of labor in society in the coming decades.

And it hinges on how digitized knowledge and information can help people acquire new skills.

History is marked by two industrial revolutions which brought about the following changes: the first industrial revolution spread from central England to the entire European continent between the 1860s and the 1850s and then affected North America. During this time, human production gradually shifted to new manufacturing processes. A production and technological revolution emerged with the replacement of human labor by machines and the replacement of manual production in individual workshops by large-scale factory production. The invention and use of machines became the hallmark of this era. Historians therefore refer to this era as the machine age, and machines became the most important factor of production in this era. "Electrification" was the hallmark of the second industrial revolution, which took place in the second half of the nineteenth century. When social production and industrial capitalism developed, entrepreneurs became not only a new group but also the most important factor of production. In the era of industrial capitalism, financial innovations such as the modern joint-stock limited company, the modern stock market, modern commercial banks, and investment banks were created. At the same time, the ownership and management of capital were separated. In addition, the modern corporate system was created.

Since the development of information technology that began in the 1980s, the third industrial revolution, represented by wireless communications, computers, and the Internet, has not only brought about a new era of industrial revolution for mankind but also a series of innovations in the financial system, such as venture capital, private equity, and hedge funds, among others. The new technological revolution, represented by technologies such as cloud computing, artificial intelligence, and blockchain, has become the pedestal for the economic development of this era. At the same time, financial instruments, represented by asset securitization and venture capital, are transforming all things external into tradable assets. In this process, people's ability to access knowledge and perceive the world is being expanded, as well as information and data becoming the most central factors of production in the digital economy. This is the biggest change of all. "Digital knowledge and information as a key factor of production" is mentioned in the G20 Initiative on Development and Cooperation in the Digital Economy. Human society is entering the digital economy at a rapid pace. And each individual is entering an era of lifelong learning and growth and a new knowledge and information explosion.

Not only has information and knowledge brought about an explosion of knowledge, but the development of the information economy has also challenged traditional economic thinking. Buchanan, the Nobel Prize winner in economics, has said that modern economics "has lost its passion for salvation and its dream of fairness." We need new economics to discuss the underlying logic of the digital economy and, in particular, of information development. One of the most important reflections of the last two centuries has been on the paradox of Western economic theory about the "invisible hand" and the "self-interest that produces altruistic results." The digital economy is built on the assumption of shared qualities rather than resource scarcity.

How, therefore, can we build an economic idea based on mutual benefit? This is an important challenge for our research in digital economics.

Second, digital technologies have become key to the transformation of productivity and production relations. Artificial intelligence and blockchain technology are to be highlighted. Although the widespread adoption of AI technology is still roughly decades away, we are seeing economic and social changes arising from AI technology taking place. For example, automation is replacing more and more people's jobs, the rise of freelancing, and the transformation of the labor market. At the same time, blockchain technology is not only bringing about a new understanding of machine credit and technological contracts but also offers solutions for creating a "trustworthy" production relationship and solving the problem of income distribution in business organizations. Thus, we see that the future of the economy in the digital economy depends to a large extent on the social and economic changes brought about by the application of new technologies.

Third, digital literacy becomes the most important quality for each individual when facing an uncertain world. Here we are concerned with the formation of a cognitive framework for a complex world. The possession of digital literacy becomes a basic requirement for every individual, determined by the knowledge economy that has been shaped by the digital age. The EU identifies digital literacy as a core citizenship quality. It proposes a framework of five domains of digital literacy, including the information domain, the communication domain, the content creation domain, the security awareness domain, and the problem-solving domain. And I believe that digital literacy is essential as a basic competency for building a cognitive framework for a complex world. Only digitally literate individuals can quickly build new frameworks and approaches to cognition in the digital world through an interdisciplinary approach.

In the new era of the digital economy, we are seeing the development and change of the digital economy, especially in developing countries represented by China where technology is the most important basis for achieving accelerated economic development. As China's economy enters a new normal, the development of the digital economy has also become an important element of supply-side structural reform. For companies, the development of how to be driven by the integration of the real and digital economy becomes an important challenge. And for entrepreneurs and innovators, how do they face the new challenges of the digital economy? Especially in the light of the new concepts and ideas that are constantly emerging in the field of information and knowledge, the establishment of a set of cognitive frameworks for understanding the digital economy is an issue that needs to be addressed urgently. In the digital economy, there are different rules of development than in the industrial economy, with digital services and products having zero marginal costs, heterogeneous effects, and sharing models. How can the underlying logic of the new economic phenomenon brought about by new technologies be understood? How can a new economic model based on technological innovation and entrepreneurial thinking be constructed? How can an overall theoretical framework for the development of the digital economy based on Chinese reality be constructed?

These are the important questions I need to think about at the beginning of building a theory of digital economics.

Business Enlightenment for a New Era

Once the following two contexts for the emergence of digital economics are understood, with its inclusion of global financial capitalism and the development of the digital economy, the content framework for the study of digital economics and the fundamental questions to be addressed by this body of theory will be specifically explored. In "Principles of Digital Economics," the discussion of the content of digital economics is divided into three chapters.

In the first part, theories about technological innovation, economic growth, and entrepreneurship-related theories involved in digital economics are initially explored in the context of the economics of the information world. It focuses on the differences between digital economics and traditional economics based on the logic of technological innovation and technological development and is unpacked from economic thinking. An evolutionary and emergent-based economic thinking is established through the introduction of theories related to complex economics. This is not only the most fundamental methodological approach to digital economics in the research process but it is also a deeper exploration of digital economics in conjunction with the theories on interdisciplinary research in complexity economics. In other words, the inner logic of digital economic growth, a set of models of digital economic growth based around the construction of entrepreneurial innovation, and the theory of incremental payoffs are highlighted in this chapter.

The second part, Value Network Theory focuses on value theory, property rights theory, and firm theory in digital economics. Through the study of traditional and Austrian economic thought, the differences and commonalities between these areas of digital economics are distilled, and a paradigm of understanding is developed that incorporates the results of cutting-edge economic thought and builds on it. The focus of this chapter is on the economics of consensus communities and complex networks in the theory of business organization by introducing a discussion of blockchain technology, comparing and analyzing the network organizational structure created by blockchain with that of business organization, and based on this, exploring cryptonomics and the business ecology theory of network organizations. It is by considering the network organization as a subject with a dual structure of both market and enterprise that the theory of digital economics is given a new paradigm for understanding the relationship between enterprise and market, and one that allows for a deeper understanding of the basic economic principles in the field of digital economy. In this chapter, we focus on the interpretation of technological innovation, organizational change, and theories of property rights and value, which have received less attention in traditional economic theories, from the perspective of digital economics, and introduce ideas from nonmainstream economics, such as new institutional economics, new structuralist economics, and Austrian economics, into the field of digital economy for discussion and research, and construct a "value network" based. This paper will discuss and study the ideas of neo-institutional economics, neo-structuralist economics, Austrian economics, and other nonmainstream economics in the field of digital economy and construct a theory of market and enterprise based on "value networks" from the perspective of digital economics.

In the third part, the monetary theory, institutional economics, and capital theory of digital economics are mainly discussed in the capital theory of the digital economy. As blockchain technology brings about new changes in production relations, Internet technology brings about a redistribution of information-based economic factors, and artificial intelligence brings about changes in productivity, it is important to build a theory of capital that is applicable to understanding the field of digital economy. Unlike the research on capital theory in traditional economics, capital theory in the digital economy focuses on the study of niche economic ecology, i.e. through the study of blockchain, artificial intelligence, and internet technology, the monetary theory, institutional economics, and capital theory originally used in macroeconomics are applied to the study of industry and enterprise ecology, so as to obtain a more practical and innovative capital theory. The key to this part of the study is how to use theories of macroeconomics, institutional economics, and capital to study industries and the ecology of firms. The key to this part is how to understand the business organization as a niche economy based on consensus networks and then to build economic institutions for such a niche economy and to use monetary policy and institutional economics to build the strategy of the business organization, which is why the future business organization needs a "chief economist" more than a "chief strategy officer." This is the reason why organizations in the future will need a "chief economist" more than a "chief strategy officer." In other words, in this chapter we focus not only on the economics of money and capital but also on the changes and applications of monetary and financial theory in the digital economy as a result of technological innovation, especially the analysis of the practical and industrial theories of business represented by the phenomenon of the pass-through economy.

In short, the aim of the theory of digital economics is to combine cutting-edge economic thinking with traditional economic theory in the specific field of the digital economy on the one hand. It is an attempt to explore a set of frameworks for understanding the digital economy and to help entrepreneurs and innovators to build a deeper and more ambitious picture of the future development of the digital world, as well as a more valuable methodology in practice. On the other hand, digital economics is also an exploration of innovative economic theory, attempting to build a logical and explanatory economic framework by integrating knowledge from economics, computer science, sociology, finance, management, political economy, and political philosophy. I think this is a meaningful attempt by a scholar in this era, and it is also the responsibility and courage that a scholar in a large digital economy should have.

In my previous monographs "Boundless: Cognitive Upgrading in the Age of Artificial Intelligence" and "Origins: From the Turing Test to Blockchain Consensus," my research focuses on the history of the development of technological ideas and the construction of an interdisciplinary body of knowledge to build a cognitive framework for a complex world. In "Principles of Digital Economics," I hope to build a framework and ideas for innovators to solve fundamental problems by examining new problems in the field of economics.

2018 marks the fortieth year of reform and opening up. During these four decades, China has implemented extensive and deep reforms in the economic sphere and has achieved remarkable results worldwide. Its average annual growth rate has reached 9.4% and it has become the second-largest economy accounting for more than 15% of the world's GDP. From the perspective of the history of human civilization, China's reform and opening up is a landmark. Never before in history has an economy with such a large population achieved such high rates of economic growth over such a long period of time. On the one hand, the socialist market economy system has been implemented by learning from the Western market economic system and adapting it to China's national conditions. Moreover, resource factors are allocated through the market mechanism. On the other hand, we are constantly guiding China's economic development through theoretical innovation and pragmatism. China is becoming a global leader. Chinese economics has also undergone a process of changing its thinking and theories from learning from the West to proposing theories that are in line with its own practice. This is also the historical mission and responsibility of global power. As a result, economists with international perspectives and Chinese ideas such as Li Yining, Yang Xiaokai, Lin Yifu, and Zhang Weiying have been born one after another.

I am sure there are many shortcomings in this book due to my lack of knowledge, so I would appreciate your advice. This is an original discipline that has only recently come into being. Therefore, there is still a lot of work to be done. I will continue to iterate and update the results of this field through my tireless academic and practical work in the future so that I can contribute to the development of this field to the best of my ability during my lifetime. I conclude the foreword with a paragraph from the end of the book.

In the future, I will continue to publish books related to this field, so that the theory of digital economics can be gradually improved and theoretical research results with a global perspective and Chinese characteristics can be established. Karl Heinrich Marx once said, "If we have chosen a profession in which we can best work for the welfare of mankind, the burden must not overwhelm us. For it is a sacrifice for all. What we feel then is not a poor, limited and selfish pleasure. Our happiness will belong to millions of people. Our cause, though obscure, will be eternally useful. And when noble people face our ashes, they will shed tears." What I am engaged in may not be the greatest of causes, but this cause is a goal that an academic would like to devote all of his life to in his short life. Although digital economics is in its toddler stage, I believe it is a long way to go, and if we persevere, we will succeed. Digital economics is an emerging discipline. We also welcome all scholars and experts to study it and walk with us.

This is used as a preface to the main text.

Shanghai, China
October 2018

Zhiyi Liu

Contents

Part I Economics of the Information World

1 Introduction to Digital Economic Thoughts 3
Information Technology and Economic Development 4
The New Order of Digital Economics 7
The Spiritual Connotation of Consensus Thought 8

2 Digital Economic Technology and Innovation 13
The Peak of the Digital Economy Wave 13
Computer and Internet Revolution 15
Digital Economy Technology and Innovation 17

3 Growth Theory of Digital Economics 23
The Equation of Digital Economic Growth 23
Limitations of Traditional Growth Theory 26
A New Perspective of Digital Economics 28

4 Complexity Economic and Order of Evolution 31
"Principles of Economics" and the Continuity Hypothesis 32
Complexity Thoughts Based on Emergence and Evolution 33
Complexity Economics and the Principle of Increasing Returns 36

5 Technological Elements and Enterprise Innovation 39
Schumpeter's Theory of Enterprise Innovation 39
Technological Evolution Theory and Economic Structure 42
Entrepreneur Theory and Innovation Thoughts 46

Part II Value Network Theory

6 Value Theory of Digital Economic 51
Traditional Economic Value Theory 52
Austrian Economic Value Theory 53
Digital Economic Value Theory 55

7 **Economic Thinking of Blockchain** . . . 63
Open Network and Encryption Economy . . . 63
Communities of Consensus and Complex Networks . . . 66
Platform and Business Ecology Theory . . . 69

8 **Blockchain and Austrian Economics** . . . 73
Origin of the Austrian School of Economics . . . 74
Methodology of the Austrian School of Economics . . . 76
Algorithmic Economics and Austrian Economics . . . 78

9 **Network Organization and Innovation Management** . . . 83
Dual Structure of Networked Organizations . . . 84
Network Platforms and Ecological Construction . . . 86
Innovation Strategies for Networked Organizations . . . 89

10 **Property Rights and Equilibrium Theory in Digital Economics** . . . 93
Equilibrium Theory of Digital Economics . . . 93
The Property Rights Theory of Digital Economics . . . 98
Blockchain Networks and Governance Mechanisms . . . 101

Part III The Capital Theory of the Digital Economy

11 **Monetary Theory of Digital Economy** . . . 107
Monetary Theory of Traditional Economics . . . 108
Monetary Theory of Austrian Economics . . . 109
Niche Economies and Token Economy . . . 112

12 **Institutional Economics in the Digital Era** . . . 119
Economic and Social Institutional Changes . . . 119
Smart Contracts and Coase Theorem . . . 122
Institutional Change and the Digital Economy . . . 125

13 **Money Supply and Demand and the Ecology of the Token Economy** . . . 131
Money Supply and Demand and the Credit Medium . . . 131
The Principles of Economic Eco-design . . . 135
The Business Logic of the Token Economy . . . 137

14 **Capitalism from a Digital Economics Perspective** . . . 143
Dispute of Two Cambridge and Complexity Paradigm . . . 144
Capitalism of the Austrian School of Economics . . . 146
Capitalism and Innovation in the Digital Economy . . . 149

15 **Economic Order in Open Network** . . . 155
Transaction Network and Capital Structure . . . 156
Open Economy and Monetary Ecology . . . 158
Token Economics and Price Theory . . . 161

References . . . 167

Part I
Economics of the Information World

Chapter 1
Introduction to Digital Economic Thoughts

After publishing a series of articles focusing on the technical thinking of the digital economy, especially after the completion of the book "Smart Economy: Understanding the World Using Digital Economic Thoughts", I realized that I had to extend the research field to the entire field of digital economics. There are three basic starting points for research in this field:

First, the study on information technology paradigm and technical thinking is essentially the study on the evolution of technical paradigm since the invention of computers, and the area where technical evolution brings the greatest changes is the economic field. Whether it is computer, artificial intelligence, the Internet, big data, the Internet of Things or blockchain, the obvious changes brought about by changes in different technical paradigms lie in economic and social fields. The shift from study on technical thoughts to that on digital economy is a process that happens naturally, and it is a perspective transformation from the bottom-level philosophical speculation and caring for human civilization to the actual economic phenomena and future development trends.

Second, the three core concepts of "information, order and consensus" summarized in the study on the paradigm of information technology are related to economics to varying degrees. Information describes the concept of technologically biting the physical world, which arouses my interest in exploring the digital world through economics. The concept of order focuses on how to understand the formation of a new economic order in the world of digital economy. The concept of consensus needs to discuss how to reach a new consensus in the process of human collaboration, and what kind of consensus can promote human civilization to evolve into an economic ecology that meets the needs of civilization development and satisfies individual subjective values and feelings. Through the study of these core concepts, an important internal logical connection between technical thinking and digital economics as well as the basic concepts of digital economics have been constructed.

Third, nowadays when digital economy research has been receiving more and more attention, most digital economy studies focus on the impact of digital technology on economic development and traditional industries, while there are few

Z. Liu, *Principles of Digital Economics*, Contributions to Economics,
https://doi.org/10.1007/978-981-16-9020-4_1

systematic study on the internal logic of digital economy development through economics. The study on digital economics, on the one hand, explores the digital economy-related topics of economic significance through the analysis of cutting-edge economic thoughts in the technical field, while on the other hand, introducing research findings in information technology science, sociology, and anthropology through an interdisciplinary approach to systematically discuss the core of new economics thought in the era of digital economy.

Actually, the above issues are closely related to the theoretical issues facing by the mainstream economics represented by neoclassical economics as has developed so far, which I will discuss briefly here.

As is well-known, the core of discussion by the current mainstream economics, that is, neoclassical economics, lies mainly in resource allocation, to be accurate, the price issues. As Zhang Wuchang, an economist said, the current mainstream economics (including macroeconomics and microeconomics) should be merged into a "price theory", which is kind of extreme to the mainstream economics taught in schools but do be justified. In the era of digital economy we are now facing, as the new technologies such as the Internet, artificial intelligence, and blockchain continue to develop, the challenge posed on us is to establish a new economic theoretical framework to adapt to the changes in the economy of the real world. What I have to do is to develop a set of economic thoughts fitting the digital economy era by borrowing the current frontier research results of economics. I will integrate new economics thoughts into the system of "Information, Order, and Consensus" I wrote earlier to form a set of logical and consistent academic theories. In the next step, I will divide this topic into three parts for discussion: information technology and economic development, the new order of digital economics and the spiritual connotation of consensus thinking.

Information Technology and Economic Development

Firstly, let's have a discussion on the concepts related to "information". In the narrow sense, the concept of information we discuss here covers any object that can be digitized while what the digital economics studies is the development law and internal logic of economics in the digital world. In the broad sense, information has developed from a mere media object to the main body of the digital economy due to technical development and evolution, and the scope of economics has also extended from the natural world to the virtual world, and from the traditional industrial economy to the digital economy.

During this evolution, three basic questions about information have been raised:

First, the connotation and extension of information. I had specifically discussed the importance of information in the development of human civilization in my previous articles. As the information technology revolution develops, information has be raised to be the basic element of civilization from the margins of civilization, during which not only the development of information technology and information economy but also changes in human's way to know the world have been resulted in. As we see that whether it is McLuhan, a master in the media field, or Kevin Kelly and Negroponte, thinkers in the Internet field, they have been studying based on the technology and ideas with "information" as the main body; "information" has become an important perspective for understanding technical thoughts, and also the starting point for us to determine the development of human civilization in the future. Therefore, it is necessary to introduce the philosophy of information technology into the digital economics for discussion.

Second, in a narrow sense, the evolution of information technology includes but not limited to the progress from Turing's invention of computers to artificial intelligence, big data, blockchain, etc.; in a broad sense, it covers the from languages, writing to electronic media and Internet media. The essential problem needed to solve in the discussion on the evolution of information technology is the law and logic of economic development in the technical development and evolution. We can think that in the current economic development, technology is a more important element than price and resource while the development of information technology relies on a major change in the perception of the outside world. In short, the evolution of technology can be explained through the economic evolution, while the evolution of technology depends on the interpretation of changes in the paradigm of thought. Therefore, we need to discuss the issue of technical evolution while exploring into the underlying ideas about specific technologies.

Third, network economy is an important theme of the information economy. Note that here it refers to the "networked" economic paradigm, not just the Internet. If we say the industrialized economy is a homogeneous competition represented by pure competition, then the network economy can be said a heterogeneous competition represented by the monopoly economy paradigm. Here, the network economy mentioned not only involves the development of the Internet and blockchain technologies, but also covers all specific network and non-network phenomena through the dimension of network, that is, consider the network as an underlying concept for economics studies same as the economic concepts such as enterprise and market. In this sense, neo-classical economics is the study on network economy with incomplete structure, while digital economics studies the basic logic of network concepts in all different transactions. Therefore, we specifically discuss the "value network theory" in the second part to study "network" in the economics in a broad sense. Digital economics is also constructed in the networked world and networked market theory.

One of the core issues of economic research is how to grow, around which has formed economic growth and cycle theories, including Adam Smith's theory of social division of labor, Malthus's population determinism, Schumpeter's innovation theory, etc. But Harold-Domar model [1] merged, marking the starting point of modern growth theory, while the emergence of the Solow model [2] indicating the maturity of modern growth theory. The Solow model proposes that economic growth is determined by population growth rate, savings rate, and technical level; however, technical level (total factor productivity (TFP)) is an exogenous variable rather than an endogenous variable, so it fails to explain the true source of long-term economic growth. In addition, classical economics has failed to explain the long-term economic growth of various countries and the huge differences in per capita incomes in different countries all the same.

In the afternoon of October 8, 2018, the Nobel Prize Committee awarded Nobel Prize in Economics that year to William Nordhaus and Paul Romer, economists from the United States for their studies on innovation, climate and economic growth. The former is a scholar at Yale University, who focused on the economics of climate change. The latter is the former chief economist of the World Bank, who created the famous "endogenous growth theory." This award is not only for the theories of "development economics" and "environmental economics", but also bringing re-understanding and thinking about the basic logic of human economic growth. In particular, Romer's endogenous growth theory has inspired me a lot in my study on the growth issues in the field of digital economy, and also made me systematically think about the growth theory in the development of economics. Romer, the winning economist proposed in the mid-1980s a new economic growth theory with "endogenous technical change" as the core [3], while establishing a long-term growth model with endogenous technical change as the core, believing that technical progress is the core of economic growth, thereby proposing that innovation makes knowledge a commodity, and knowledge and human capital have an effect of increasing returns to scale, so as to promote long-term economic growth. This theory is one of the basic logics we follow to study on the development of the digital economy, that is, to understand the growth issues in the digital economy through study on technology.

In fact, in the long history of human development, economic growth would not steady and constant. During the 1800 years of the past 2000 years, the human economy had hardly grown, until the Industrial Revolution that the first rapid growth occurred. In the 1950s, the information technology revolution represented by the computers and the Internet made the human economy grow in a vertical straight line, that is, an explosive exponential growth. The core of the "endogenous growth model" created by Romer is that "knowledge" is an element of economic growth. Knowledge accumulation will affect the workforce level and the rate of return on capital in an endogenous manner, while innovation can make knowledge a commodity, creating the monopoly theory. His theory is one of the starting points for us to study on digital economy, and what we need to study is economic phenomena in the field of digital economy based on the "technical endogenous theory" [4].

In summary, the first important part of digital economy thinking to be studied is how to combine new economic thinking with new technical paradigms. I will study new economic paradigms such as information economics, complex economics and network economics to study the characteristics of new economic elements. I hope that I can provide a thinking dimension completely different from the neoclassical economic paradigm to the readers to understand economy by introducing the current theoretical research on these three fields in economics.

The New Order of Digital Economics

We see that as time goes, the mainstream economics develops into a kind of economics dominated by study on mathematical economic model gradually, while many aspects of economic research, including politics, history, sociology have been neglected. Actually, these contents constitute a theoretical system with a long tradition in the development of economics. One of my tasks is to go beyond the current neoclassical economics paradigm, which takes mathematics as the main research method, to study the essence of economics at a higher level through interdisciplinary methods (including political science, sociology, psychology, anthropology) and so on, to build a new economic thought that is closer to the essence. Let's discuss the important issue to be studied in economics in the digital economy era, that is, how to establish a new economic "order".

Firstly, let's have a discussion on the nature of economic research. I believe that economics is actually the study on human behavior. In the book "The World Economy: A Millennial Perspective" [5] by Maddison, a famous scholar, he described the development of economic history in the past 2 years in Europe, including the evolution of population history, changes in social technology and so on. He concluded three basic dimensions of economic development: First, human's control and colonization of nature, that is, the transformation of nature. We have seen that whether China's transformation of the area south of the Yangtze River or the Europeans' transformation of the American continent, they had promoted the economy to develop tremendously. The second is trade and circulation of capital. We have seen typical cases of economic development in ancient times, such as those in Venice and the Middle East, all of which were brought about by their advantages in the trade and capital circulation, the most important economic development advantages before the modern industrial revolution. Third, innovation in the fields of technology and systems; for example, the UK's advantages in system innovation and navigation technology cultivated the glory of Great Britain, while the technology innovation and venture capital system of Silicon Valley in the United States incubated a revolution in Internet technology. Undoubtedly, in the digital economy era, the third element is what we need to discuss most. Before we discussed technical innovation and now the system is what we need to discuss, that is, the establishment of "order".

Actually, the concept of "order" here refers to the factors that affect human behavior. We see that the order usually mentioned in economics works includes three basic order dimensions: The first is the "order of things", that is to examine how order is formed from the perspective of technology and material. The second is "social order", that is to examine whether the order is possible from the perspective of the relationship between people. The third is "spiritual order", that is to discuss people's inner spiritual demands.

Therefore, what we have to do is not only the discussion on the study related to classical economics, but also the establishment of a set of academic ideas to understand how the economy and society to operate in a systematic manner through reference to other fields such as sociology, psychology, philosophy, and anthropology which is helpful for us to create a set of effective and inclusive theories while understanding new technologies in the field of digital economy. Besides, at this stage, we need to focus on providing other important concepts such as morality, happiness, and natural ecological harmony to the new order, so as to ensure that the value system exclusive of utilitarianism are recognized by innovators in the digital economy era to promote the reshaping of social value and economic order. Especially in the economics research in the digital economy era, discussion on the underlying values and the morality of wealth should never be absent. Only by establishing proper order and values, can we provide the communities with the correct "relationships of production", and can the social and economic changes be promoted to a greater extent through informatization.

The Spiritual Connotation of Consensus Thought

Consensus building is not a new concept. Consensus already existed as early as when human beings started group life. The consensus mechanism discusses how to make a group of diversified values collaborate and make decisions in competition. In the consensus mechanism, all members in the network accept the constraints from laws, rules, and contracts, as well as recognizing the institutions that implement these laws; they recognize the identities and share the common values which is the basis for consensus mechanism to be achieved.

In addition to the discussions on technical and economic aspects, in the future we will discuss consensus that are more underlying and without which the economics would not have been established. Most researchers have been focusing on the many technical mechanisms in the consensus mechanisms of blockchain technology as well as the technical concept of the Internet of Value formed, while what I have been concerned about is what kind of consensus spirit we should establish. Here we need to have a discussion on some basic concepts concerning ethics and political economy such as happiness, justice, fairness and freedom.

At the beginning, we will discuss the concept concerning moral philosophy. In fact, good in economics means "good" in moral philosophy, and the goods in economics share the commodity attribute of "goodness", so we have reason to believe that actually, economics is a science pursuing "goodness" and "happiness" at the moral level, rather than the mere knowledge of wealth and efficiency as we have seen. There have been important works on the study of human morality, such as Aristotle's "The Nicomachean Ethics" [6], Hume's "An Enquiry Concerning the Principles of Morals" [7], Moore's "A System of Ethics", and the "Protestant Ethics and the Spirit of Capitalism" we discussed above. Only after having clarified the nature and essence of morality can we understand various decisions made by people in economic behavior, and thus breaking with the concept of "rational man supposition" proposed in neoclassical economics. In this chapter, we will also discuss the concept of happiness, because modern people have been increasingly pursuing happiness; however, in fact, happiness is not the simple result of wealth effect, but the outcome of moral choice. In the era of digital economy, we should sort out people's behavior from human nature and morality, rather than stay studying theories related to numbers and wealth in economics.

After that, we will discuss the theory of value. We will have a discussion on the study of the history of value theory in the book "The Theory of Value before Adam Smith" [8] by Ceval, a famous scholar. Of course, we will also discuss the elaboration of value from Hegel and Adam Smith as well as John Locke, et al. In fact, as we see that neoclassical economics has shifted people's focus from value to price, resulting the focus has been placed on "cost" rather than "value" in the modern economic research. In the digital economy era, more attention should be paid to the value of things rather than costs. Different people see different values in the same thing and that's why new economic forms such as the sharing economy is possible to occur. In the second chapter, we will specifically discuss the differences and continuities between the axiologies of digital economics, traditional economics and Austrian school of economics to help you understand the contents in this part.

Finally, we will discuss the theory of political economy, that is, how to distribute to achieve both efficiency and justice. We know happiness sources from efficiency, justice and freedom. We will discuss freedom when we come to the field of spiritual order, and efficiency and justice are at the core of current political economy research. We will have a discussion on the theories of Buchanan, Habermas, and Stigler on these two topics. More importantly, we will discuss the theory of happiness in depth according to discussion on the theory of political economy. As a social science, one of the basic purposes of economics is to solve the problem of happiness for human. In the digital economy, the smart contract-based technical contracts and consensus-based social networks have been based on to solve the problem of happiness, that is, to reach contracts and consensus through technical methods to form the basic values and basic contracts in the community which is also a thing blockchain technology should do in the future.

In summary, actually the consensus spirit we discussed solves the moral problem first, that is, the basic motives and needs of strangers when they get along. Then we should define the value issues in the process of applying technology to allow us to look at the digital economy in a new way. Finally, we shall realize the happiness brought about by economics, that is, the issue of efficiency and justice. All knowledge should bring happiness to mankind as a whole which is one of the basic starting points in our research.

These three parts above are my brief introduction to the basic ideological framework of digital economics. In fact, most of the content involved in this framework is the problems have been studying on sine the economic thoughts generated although many have been marginalized. In fact, these issues should not be neglected in the development of technology, but should be elaborated more systematically and in-depth.

For conclusion, I have three main characteristics in the study on digital economy which I hope to be well-known to readers:

Firstly, start related work based on the study on "people", the essence of economics rather than focus only on mathematical analysis or exploration of commercial value. Economics, as the essence of social science research, has been neglected as the neoclassical economy and the technology is rising. Part of the academic thoughts in this part comes from the economic research methods of the Austrian School of Economics which are based on human behavior, so in part of the book, we introduce the logic of the thoughts of the Austrian School of Economics specifically. I believe that all knowledge should be based on "human". Human nature has varied little since millions of years ago and it and the sociological study of different people living together (especially strangers) should not be neglected no matter how society and technology progress. Therefore, we will attach great importance to the relationship between technology and people, the relationship between technology and the community, and the relationship between technology and the spiritual order of people to build digital economics theoretically based on human behavior, rationality, and how to divide labor and collaborate.

Secondly, discuss the most cutting-edge economic theories and interpret the classics, for example, the impact of technology on the economy, especially the impact of the development of information technology on the economy. The way we discuss is to sort out the theories involved in the new digital economics according to studies on information economics, social economics, network economics, and complex economics, etc. I would not like to invent new academic concepts, believing that any new things come from original things after combination and evolution. So in research on the digital economy, we should also study on new economics and inherit traditional economic theories rather than create many specious new concepts for research.

Thirdly, in my academic research, an important driving force is my concerns about the happiness of the human race and the formation of innovators' knowledge and pattern. So, I try to help people develop a way to observe the world based on interdisciplinary thinking. "Unbounded: Cognitive Upgrade in the Era of Artificial Intelligence", the first monograph I wrote, discusses how to improve intellectual knowledge. Being a scholar, I believe that the truth is not unique, but we can build a more solid and powerful ability to cognize through systematic learning of knowledge.

In short, behind the key word of information in this book, subject ideas related to information technology are discussed; the order focuses on the ideas of related disciplines such as economics, sociology, and psychology, while the consensus discusses thinking concerning philosophy, politics and ethics. I would like to contribute to your establishment of a set of thinking frameworks to think and observe independently of the world by borrowing ideas of the above subjects.

The above is what I want to introduce concerning the basic thought research of digital economics, and they are also the main points of this book. It is worth noting that since we have adopted a continuous iterative approach to discuss digital economics, more improvements and supplements will follow, and more cutting-edge ideas will be introduced and discussed. Therefore, I hope that readers are able to perceive the essence and framework of this article, rather than focus on certain knowledge modules. Next, we will start to build the digital economics theories step by step, hoping it will contribute to your thinking.

Chapter 2
Digital Economic Technology and Innovation

In this chapter, we will have a discussion on the historical relationship between the digital economy and the innovation. In fact, we have been keen to the discussion on the digital economy since it represents the important achievements in the revolution and innovation of information technology represented by the computer, the Internet, and the Artificial Intelligence in the past 70 years. The information revolution has been in so fast a pace that it has produced great influence on human society and the economy.

So, today we will discuss on the history of digital economy and innovation as well as the theory of innovation based on digital economy. By discussing herein, we know that the key to the digital economy is a technological revolution of turning the real world into bits and information based on the ideas of mathematics and algorithms. In addition, we will discuss the difference between this round of revolution and the previous technical innovations.

We will divide it into three parts for discussion: the peak of the digital economy wave, the revolution of computers and the Internet, and the digital economic technology and innovation.

The Peak of the Digital Economy Wave

The representative technology of the digital economy we often discuss broadly is the information technology represented by communications and computer-related technologies, and in a narrow sense, the technologies concerning the Internet, cloud computing, big data, blockchain, Internet of Things and the Artificial Intelligence. However, no matter how to define it, we conclude it as an important part of the fourth industrial revolution for which the most important technologies are the computer and Internet information technology, as well as the Artificial Intelligence and blockchain, Internet of Things, and quantum computing technologies, etc. based on them.

Z. Liu, *Principles of Digital Economics*, Contributions to Economics,
https://doi.org/10.1007/978-981-16-9020-4_2

The fourth industrial revolution has brought about huge impacts, especially the impact of the digitization wave on the manufacturing industry in the industrial economy, as well as on personal lifestyles. We discuss this issue from three perspectives:

Firstly, in terms of business logic, the wave of digitization in different industries has deeply changed the business logic and operating methods in traditional industries, especially the artificial intelligence technology bringing about the automation revolution and the Internet technology has led to the changes in people's behavior and methods. We see that the digital economy is bringing about changes in business logic with C2B, sharing economy and experience as the core. On the one hand, manufacturing enterprises are improving their survival efficiency based on fully automated intelligent technology, while on the other hand, they integrate their supply chains through the big data technology to meet the personalized needs of individual users. The Internet, blockchain, etc. have been disintermediating in a decentralized manner to improve people's collaboration efficiency, reduce market transaction costs, and establish a market mechanism based on technical contracts.

Secondly, in terms of organizational ecology, digital technology has allowed the new business ecological models to be constructed constantly, especially blockchain technology has brought important changes to decentralized organizations. In the industrial economy era, enterprises improve the efficiency of internal collaboration and reduce communication costs using digital technology, making the organization flat and lean. On the other hand, the digital technology represented by the Internet and blockchain technology causes the boundaries of the organization to disappear. In future, organizations will be based on open platforms and business ecosystems. Business organizations will promote the self-motivating community construction and efficient collaboration within the ecosystem through the construction of business ecosystem of digital economy which, on the one hand, improves the organization's response to the rapidly changing needs of the market and customers, on the other hand, enhances its competitiveness through integration of ecological resource.

Furthermore, digital technology has been reshaping the people's values of work and life. As the technologies such as artificial intelligence, the Internet, and blockchain develop, huge numbers of indigenous people of the digital age had been given birth to. In the Internet community, they grew up in the cultural soil of openness, freedom, democracy, equality, sharing, transparency, and diversified values, enjoying cultural genes completely different from previous generations. This has resulted in their preference to the working culture of the digital age such as co-creation, iteration, flattening, unitization, and crowdsourcing, and their pursuit for a sense of value and belonging at work while pursuing individual life and the value of self-realization.

Above is the discussion we have based on the current situation of the digital economy wave. We should know that we have been experiencing the fourth major industrial and social transformation since the first industrial revolution in the eighteenth century which is mainly characterized as the integration of physical space, cyberspace and biological space brought about by a lot of technological innovations. Its speed in technological development and diffusion as well as its depth and breadth of its impact on the human society are far beyond the three industrial revolutions. Therefore, our society needs to be prepared in all aspects. Digital economics is a study of the economic thinking based on digital technology in the fourth industrial revolution. Anyway, I want to point out that it can not explain problems in all fields.

Computer and Internet Revolution

After discussed the topic of digital economy and innovation, we will go back into the historical process to have a look at how this technical revolution was developing. Due to space limitations, we mainly discuss how the computer and Internet technology develop; to be more precise, discuss how the computer and Internet revolution happened in the United States. Although many benefits have been brought about by the computer and Internet revolution, there are still big problems, which mainly concern three aspects:

Firstly, the information technology develops, exacerbating the phenomenon of inequality. Since the information technology was promoted in different paces, regions and families with better economic conditions benefit more quickly, resulting in inequality in the information education as well as in understanding of information technology.

Secondly, the information technology develops, giving birth to the outbreak of data security and data privacy issues. Besides, it is the few monopolistic Internet companies that master a large amount of data, resulting in huge security and privacy issues of Facebook and the inappropriate use of user privacy data which had contributed to the introduction of the General Data Protection Regulation (GDPR) [9] in Europe.

Moreover, the Internet has brought too much information, causing distraction to people, the proliferation of popular entertainment thinking, and the education and friendship problems of teenagers. No doubt, online games, online social networking, and online violence information have great negative impact on education which can not be neglected.

Up to time now, the information technology revolution of computers and the Internet has entered into a new stage: the technological revolution based on the influence of computer algorithm ideas has given birth to the third wave of artificial intelligence technology, and the ideas based on the Internet information technology revolution have caused the blockchain technology revolution to occur. I will briefly analyze the development history and influence of the two technologies.

The concept of artificial intelligence took shape in the 1950s, and it has experienced three waves in the development: the first was the machine translation based on logical reasoning that began in the 1950s, represented by Dartmouth Conference in 1956. In this wave, people focused on discussion of the technical ideas and underlying mathematical logic of artificial intelligence. The second was the expert systems relying on knowledge accumulation to build models in the 1970s. As the Hopfiled neural network and BT training algorithm were proposed in the 1980s, various expert systems emerged and the fifth-generation computer was proposed in Japan, however, artificial intelligence was still far from entering into people's daily lives. The third was the artificial intelligence wave which has been focused on data, machine learning, deep learning and other algorithms since 2006, represented by the event that in 2006 AlphaGo defeated the Li Shishi, the human Go player. Since then, the algorithms represented by deep learning had been applied on a large scale, and artificial intelligence technology had been widely used in speech recognition, computer vision, autonomous driving, smart cities, etc.

At present, we can start our analysis from three perspectives: First, regarding to the technology, the research on general artificial intelligence technology has come to the breakthrough stage, and most of which are carried out by the top large enterprises worldwide, such as Google, Facebook and domestic BAT companies. There are more companies of dedicated artificial intelligence and more application scenarios, which are mainly driven by small and medium-sized enterprises. Second, from the perspective of application, artificial intelligence has good application prospects in fields with rich data and more pain points, such as industries of manufacturing, automotive, retail, healthcare, and finance. It is especially optimistic for its use in autonomous driving, financial risk control, and intelligence in the future. Third, the artificial intelligence has been focused on breakthroughs in the field of professional weak artificial intelligence, while the general artificial intelligence is relatively slow in pace. In the short term, we expect to deliver more personalized services to more consumers and enterprises without increasing marginal costs, which is the future of artificial intelligence technology that we can expect.

Actually, as early as 1989 when David Chaum created a digital currency system called DigiCash, the basic technology of the blockchain had been in accumulation and the real iconic event happened in November 2008, when Satoshi Nakamoto published the famous paper "Bitcoin: A Peer-to-Peer Electronic Cash System", and in January 2009, he mined the founding block using his first version of the software. We can think like this that the Bitcoin emerged because under the influence of the global financial crisis in 2008, technology geeks put forward a set of resolutions of decentralization due to distrust in the centralized organizations, showing the digital world's distrust to the economic model emerging in the capitalism world. Since Brexit on June 23, 2016, North Korea's fifth nuclear test in September 2016, and Trump's election on November 9, 2016 followed, adding more uncertainties to the mainstream economy of the world. As a result, Bitcoin, which avoids risks and serves as an alternative to the mainstream economy, began to recover. The market demand increased, and the scale of transactions expanded rapidly, bringing the rapid growth of Bitcoin in the next more than 1 year. Yes, blockchain still has many technical, commercial and legal problems and so far there have been no successful large-scale commercial applications. In other words, the blockchain is still at the earliest stage for the business ideas and technological innovation. Due to space limitations, I will conclude into three points here rather than discuss the history further:

Firstly, the revolution of computer and Internet technology is developing in a fast pace, bringing new innovations constantly. Decades are rather a short period in other industries, while for which information technology has gained such glorious achievements and this is the main reason why we are concerned about this important phenomenon of digital economy. The development of information technology is not only a most important practice of human beings to promote the society to develop relying on innovative technologies, but also a painstaking attempt of mankind to shape a civilization of a new virtual world other than the real world. This is the most important social practice of mankind after the industrial revolution.

Secondly, as the digital economy is continually developing, we see that the prices of corresponding products and services have been constantly dropping which is an important reason why it has been welcome by consumers and has gradually changed other industries. As we can see, the absolute prices in other industries have been rising, while the prices in the information or IT fields are constantly dropping, while enhanced performance and services of the products are provided (not to mention most of the Internet applications are free). This is a focus in study on digital economy, and also the basic logic of our subsequent discussion on the economic principles of increasing returns. Note that there are two characteristics of digital economy completely different from the real economy: on the one hand, the right to use in asset sharing has replaced the property relations dominated by ownership of fixed asset invested, embracing an opportunity to reconstruct the productive relations. On the other hand, according to the operation of digital assets, a balance between the efficiency and fairness in income distribution is achieved, providing a good opportunity to expand people's demand and to promote supply-side reforms.

Thirdly, in the revolution of information technology, whether the computer, the Internet, the big data, the artificial intelligence or the blockchain, all new technological revolutions base on a basic logic of turning all problems into mathematical problems. This logic is as effective in the development of the digital economy in the past few decades as in the historical cycle of IT revolution technology in the past two centuries or so.

Therefore, how to clarify the historical origins and basic logic in the development of the digital economy is the basis to our subsequent discussion on theories of digital economics, and also the key for us to understand future changes in the economic world.

Digital Economy Technology and Innovation

In the last chapter, we will discuss the digital economy technology and innovation, especially we will analyze the nature of innovation. It is generally believed that innovation is something to create new and revolutionary things which was emphasized in "From 0 to 1" by Peter Thiel targetedly. However, I think innovation is to recombine existing elements. Through the accumulation and creation of knowledge, we keep discovering the nature of things and the properties after combination, and then we apply the things after combination which is innovation.

In the past few decades, China has been firmly embedded in the global division of labor by introducing technology, capital and expanding production capacity to become the world's second largest economic power. As the international division of labor system reshapes and various elements of the Chinese economy have increased costs, China has faced great difficulties in ways to innovate by just expanding production capacity and increasing foreign investment. Besides, conflicts in international trade have also triggered China's demand for endogenous innovation. Therefore, we need to find an innovative model that fits us. We will discuss from the following aspects:

First, we discuss the important role of innovation in study on economics. Since the current mainstream economics research is to study economic growth, believing that continuous economic growth raises the quality of life for each individual (This view is undoubtedly flawed, which is also the reason to the current changes in the global economic structure), and basically, there are three ways to achieve economic growth:

First, increase and improve the input of labor and capital.

Second, give full play to professional comparative advantages through foreign trade.

Third, achieve growth by promoting innovation and entrepreneurship.

No doubt, the third path is the most important in the field of digital economy. The long-term economic growth is only be possibly achieved by innovation promoted by entrepreneurship. Later, we will discuss the influence of the ideas of Austrian School of Economics on digital economics. One of the most important points is that entrepreneurial innovation is the most important force driving digital economics which is the reason why I define myself as "innovative economist". In fact, as early as 1912, when in Schumpeter's publication of "The Theory of Economic Development" [10] "innovation" and its role in economic development were proposed for the first time, the essence of the entire economic development had been revealed, that is, entrepreneurs promote economic development by distributing the productive elements through innovation.

Then, let's give an analysis of the innovation in Silicon Valley. More than half a century ago, Silicon Valley was a nobody in the United States; however, after only a few decades, many high-tech multinational companies in the world were born here. It is a place not only gathering high-tech talents, but also giving birth to the new technological revolution. The whole world has been learning from the experience of Silicon Valley, including China, which is also building the Xiongan New District known as the "Silicon Valley of China". So, how to understand the success of Silicon Valley in innovation, and how these innovations enlighten us? The historical secret of the innovation of Silicon Valley was described in the book "A Hundred Years of Silicon Valley" written by Piero Sgaluffy, a visiting scholar at Stanford University, from which we can understand the reason to the development and innovation as well as business success in Silicon Valley:

First, the wealth was promoted to be distributed through methods of equity and options, which transformed the production relations. Silicon Valley has received so much attention from entrepreneurs all over the world because its wealth distribution mechanism does not rely on the original stock, but on increments to create and distribute wealth. This process not only brings about a huge wealth effect, but also meets people's requirements for the wealth ethics of innovators. This is the key to our understanding of the innovative spirit of Silicon Valley.

Second, there is a relatively relaxed entrepreneurial environment in Silicon Valley, and invention and entrepreneurship are extremely tolerated and encouraged, allowing the human resources to flow freely. We see that it is the success of Silicon Valley that promotes many domestic entrepreneurial industrial parks to emerge and the local policies and talent policies to change. In fact, all innovations are people-oriented and innovation can not occur without respect for the innovative spirit of entrepreneurs and a sufficiently friendly external environment for innovators.

Third, Silicon Valley has a diverse cultural atmosphere to promote innovation, that is, international talents gather there, making the Silicon Valley consensus has an international vision beyond the reach of other regions. In short, the innovation culture of Silicon Valley promotes the entrepreneurs with innovative spirit to participate in, and Silicon Valley provides a business environment for innovative companies to survive, that is, it emphasizes on individual innovative spirit and free culture and order which are the topics we will focus on when discussing digital economics.

Finally, we will discuss the relationship between innovation and technology from the perspective of economic development. As Paul Krugman, an economist said, "Productivity is not everything, but in the long run, it is almost everything." There is a direct relationship between the ability to improve people's living standards and the productivity in a nation, and it is the entrepreneur's bounden duty to increase productivity through innovation.

In other words, innovation provides the most important condition for economic growth and national prosperity, so it is very critical to improve innovation capabilities. In fact, the increase in productivity has been dropping gradually in the past three industrial revolutions, and we need to promote the productivity in the fourth industrial revolution, during which the most important thing we should do is to improve technology. After analyzing multiple inventions, innovations and technological evolution cases, Brian Arthur, the founder of complex economics, concluded in the book "The Nature of Technology" [11] that "innovation is to find something that already exists", while technology is to combine existing technologies in a new way, and innovation is the process of technical combination and evolution. Understanding this, we are able to establish the relationship between innovation, technology and economy, and know why entrepreneurial innovation is the core force driving the development of the entire digital economy.

On the other hand, the Internet economy strategically situated has impacted the industrial institutions of the traditional economy, constantly promoting the industrial transformation and business model innovation of the digital economy, which not only includes changes in way of consumer to transact, but also promotes the changes in internal economic institutions and industrial ecology of the digital economy. The innovation of the Internet economy has changed the homogeneous and standardized production and transaction modes of the traditional industry era which is promoted by the change in the information structure. The variable cost of information is almost zero, making it a new way for production side to provide customized goods and services to realize their values. In the industrial era, the economic growth model of high input, high consumption and dependence on the number of production factors has been gradually replaced by the Internet economy with a digital economy paradigm relying on the optimized combination of production factors, technological innovation, and being driven by knowledge and information. Especially, blockchain technology develops, making credit be achieved through technical contracts, and point-to-point decentralized transactions possible. For economic research, the price mechanism is no longer the whole of economic analysis. Innovative activities, especially how to promote the digital economy to develop through the incentives and constraints of the credit network, has become the focus in the study for the entire digital economy.

Finally, the innovation is not only a key issue of digital economics, but also the key to understand how the new smart economic ecology develops. We see that whether the real estate industry, the retail industry or the manufacturing industry, they have been facing digital transformation to reform the industrial chain through intelligent and networked methods to promote the optimal allocation of production factors. At the same time, through the docking with the industrial logic, the business thinking mode of the digital economy has been gradually realizing the "empowerment" of innovation in different industries. From this perspective, we may consider the digital economy an economic ecology parallel to the traditional real economy and virtual financial economy or an economic trend merging with all other economic ecology. The real economy improves productivity and allocates production factors optimally through digital transformation, while the virtual economy reduces the risk in innovation in the financial industry though digital methods, while providing more valuable and targeted services through the Internet, and solving the credit problems of the traditional financial industry through technical contracts.

In short, the core reason why the innovation of the digital economy is important lies in the borderless influence of the digital economy. The whole world has been constantly promoting the digitalization to advance, endowing unique value to the research on digital economics and continuous vitality to research on digital economics.

In summary, we have discussed three topics: the peak of the digital economy wave, the revolution of computer and the Internet, and the technology and innovation of the digital economy. Through the discussion on the information technology revolution and the history of innovation, as well as the relationship between technology and innovation, we know that the entire information technology reform develops based on constant innovation and reform and understand that the entire IT revolution is a process to virtualize and algorithmize the real world through mathematics. Then we have a discussion on the reasons why Silicon Valley has become the greatest base for innovation in the digital economy, getting to know that innovation in economic activities should be based on multiple factors concerning culture, society, and region and innovative achievements fail to be accessed by simply relying on part of the external environment changes. Finally, we discussed the relationship between innovation, technology and economy, understanding that innovation is a process of technology combination and evolution while the digital economy develops relying on entrepreneurial innovation; we also realize that attaching importance to technology and innovation is the basis to our understanding of the internal logic of the development of digital economy.

Chapter 3
Growth Theory of Digital Economics

In this chapter, we will discuss about the theory of economic growth, and study on thoughts of economic growth theory to analyze and derive the growth theory of digital economy. We see that the digital economy has been growing rapidly, showing characteristics not only of traditional economic growth, but also unique to itself. This provides us a new way to study on the growth of the digital economy: on the one hand, we must inherit the basic framework and concepts in the traditional economic growth theory, that is, link economic output with the scale of labor and capital and productivity, etc. to form a basic framework to understand economic growth. On the other hand, we should see the differences in the forms of the digital economy, especially the network economy, and explore the internal logic in the growth of the network economy based on the complicated and heterogenous research model of economic utility. Next, we will have a discussion on the growth theory of digital economics from three different perspectives: the equation of digital economic growth, the limitations of traditional growth theory, and a new perspective for digital economics.

The Equation of Digital Economic Growth

As early as the eighteenth century when the classical economics was born, we had seen that the economic growth is an important direction for economists to study on. Related scholars include Adam Smith, Malthus, and Ricardo in the classical economic period, Alfred Marshall who put forward the basic growth equation, Schumpeter who put forward the theory of technological innovation and Simon Kuznets in the early stage. Both Hamad-Domar model in modern economics and the theories involved in Rostow's "The Stages of Economic Growth" [12] were all like this.

We see an obvious academic tradition, that is, mainstream economic growth theories have been built based on general equations and production functions, that

Z. Liu, *Principles of Digital Economics*, Contributions to Economics,
https://doi.org/10.1007/978-981-16-9020-4_3

is, to find a series of variables and then explore how they affect the total economic growth in the total quantity and efficiency through functional relationships. Explaining how the entire economy develop through a framework and a theoretical system is the ideological framework of the theory of consistent growth. We put various factors such as technology, population, capital accumulation, trade, system, culture, etc. into this framework to study how these factors impact economy to develop internally. In fact, there is an important attempt of modern economic growth theory, that is, to analyze a single element of these elements, but it is still in the process to analyze multiple elements. Also, it is the focus in digital economics theory to put the multiple factors of endogenous growth in the development of digital economy into a consistent growth model for consideration.

First of all, we should understand a basic framework, that is, the output of the economy depends on the scale and productivity of labor and capital. Here, the capital includes land, other natural resources, and knowledge about science, technology and organization (in fact, knowledge and innovation are the most important capital in the digital economy system), therefore, the growth rate of an economy is deemed as a function of the change rate of these complex variables. To put it in a simply way, $Y = f(X)$, the variables in X include the capital stock, natural resources, labor of the economy, the accumulated and applied knowledge of the society, and the social environment in which the economy operates, etc. Irma Adelman, a well-known economist detailed how each variable affects the total production volume theoretically in the book "Theory of Economic Growth and Development". Here we should do nothing but know such a simple equation.

Then, we shall establish a new point above this simple equation, that is, it is not the only economic motive of humans that determine the behavioral decisions on economic growth rate and the productivity of labor and capital and economic behavior is the result in measuring material progress and other human goals. This is a important conclusion; in mainstream economics, to understand all factors as economic factors (directly, price factors) is actually a way of thinking that derive cause through results.

Both history and reality let us know that human decision-making behaviors and motives are too complicated which is determined by the complexity in human nature. So, the behaviors that lead to economic progress should not only have been motivated by economic purposes, but also by a large number of other factors. We will continue to discuss on this point later when we discuss complex economics and consensus-based political economy. We should know that economics is a science not studying on economic elements, but also studying on the behavioral decisions of humans. Therefore, more complex variables, including those concerning sociology, anthropology, psychology, history, cognitive science, etc., should be included to understand the basic driving forces and motives in economic development.

Finally, let's look back into the centuries-old history of economic development to extract the most important basic motives in economic development and raise basic questions about the theory of economic growth.

Rostow, an American economic historian and one of the pioneers of development economics, mentioned in "The Stages of Economic Growth" that, there are six basic motives that can serve as the basic motives in economic development, including:

Motivation to develop basic sciences (physics and sociology);
Motivation to apply science to economic purposes;
Motivation for active innovation;
Motivation to seek the evolution of material progress;
Motivations of consumption and of fertility and reproduction.
Based on the above mentioned, we can understand the driving forces of people's behavior in the theory of economic growth in a more systematical manner. The basic questions of economic growth theory also unfold around these motives.

Due to limited space, here we raise three basic questions: First, can long-term endogenous economic growth exist stably and sustainably? Second, in different economic cycles, what are the differences between the characteristics of economic development? Third, how to choose a path to develop for an economy?

These issues touch the essence of the theory of economic development. The traditional neoclassical endogenous theory of economic growth has a logic of balanced growth, that is, there should be a stable and unique equilibrium, serving as the premise before development. Obviously, such a theory is not in line with reality, and only being able to explain part of the phenomenons of some economies in the economic development cycle (for example, when the economies of developed countries are growing at a fast pace). The economic logic of endogenous growth we mentioned is based on the facts pointed out by Kaldor, that the marginal rate of return of capital is not diminishing. Therefore, various economic endogenous theories were born, including endogenous technological progress by Romer, the endogenous human capital by Lucas and the endogenous innovation process under innovative destruction thoughts, etc. As we see that, the endogenous elements receiving greatest attention in the digital economy are technology, innovation, and human capital and the basic ideas are derived from study on endogenous growth theory.

In short, our logic is that the analysis of the growth equation of digital economics is an analysis of complex variables, a research based on the concept of dynamic equilibrium, and a new thinking requiring combination of economic variables and non-economic variables. Seen from the development history of economic growth theory, we see that these problems reflect the essence of digital economic development: for all economic development theories, it is required to establish a corresponding relationship between equations, and for the development of digital economy, it is most important to put technology and innovation capabilities into it to distinguish it from the profit-maximizing behavior under a fixed or gradual production change function. Moreover, this equation also needs to explain the random occurrence of innovation and the inherent cyclical fluctuation behavior, etc. Therefore, we will analyze and study on the growth theory based on the thoughts on economic structure in the subsequent so as to explore the basic framework of digital economic growth.

Limitations of Traditional Growth Theory

We will have a discussion on the shortcomings of the traditional economic growth theory while exploring the full picture of the economic growth theory we have been trying to establish in this chapter.

Beforehand, we will have a glance at the mainstream macroeconomics theories, namely the ideas from Maynard Keynes, the economist influenced macroeconomic theory to the maximum extent so far with no doubt and whose theories are about macroeconomics. In the two books "A Treatise On Monetary" and "The General Theory of Employment, Interest and Money" [13], he analyzed and demonstrated the economic cycle to study on the economic cycle and economic growth. In "A Treatise On Monetary," [14] he mentioned that different decisions are made by individuals for different purposes, among which, the most important ones are savings and investment. The gap between the two triggers the economic system to expand or contract, so each economic processes have limitations in nature, producing economic cycle problems. In short, according to Keynes, investment maintains a dynamic equilibrium relationship with savings. The major discontinuous innovations therein are the force driving investment to exceed savings or to fall behind it.

Let's check the economic growth viewpoint from Austrian economists who strongly opposed Keynes for his ideas proposed in "The General Theory", claiming that with no surplus output, there is no demand and therefore the ability to consume the output of others. The owners of consumer goods do not exchange them for other consumer goods, but for tools and machines with better performance, so as to obtain more output and consumer goods of better quality in the future. Austrian economists believe that the owners of the consumer goods lend their savings to individuals dedicated for the tool and machine manufacturing(that is, innovators and entrepreneurs) by put the consumer good saved on to tools and machines, thereby promoting the birth of new products and services; more demands come only when there is better supply and increased capital loans, that is the increased innovative tools and machines are the force to expand the final consumer goods supply.

In short, Austrian school believes that real demand and real savings are increased only when the economic efficiency is improved and better supplies are created and the driving force of economics is not the demand for goods but for consumption brought about by innovation in the real world. In fact, this theory is in line with Adam Smith's thinking, since according to the logic of Adam Smith, products with higher quality and lower price are produced to open up market, thereby stimulating market demand and economic growth which the dominant Keynesianism ignores.

We see that whether Keynesianism and Austrian economics, they have been emphasizing the uncertainties in the market, but they two have gap between them. According to Keynes, in neoclassical economics, the expectation that people can judge the future accurately with the price information of the past is untrue. Therefore, it is useless to attempt to make expectations based on the mathematical probability, since the price information of the past "cannot serve as a scientific basis to form any probability calculable." Therefore, Keynes proposed that individuals predict, being dominated by "animal spirits", and form innovations based on such prediction, giving birth to business cycles and market depression. Therefore, government should intervene and guide in the formation of the market and it is the government's responsibility to stabilize investment, output and employment. Economics of Austrian school, however, while agreeing with Keynes's ideas that there is not a simple correlation mathematical relationships between past prices and future prices, believes that entrepreneurs can predict the future by mastering the subjective value of specific market participants of the past. These expectations are not numerical values but a series of subjective judgments about the market (heterogeneous expectations). Therefore, based on judgement on different viewpoints, entrepreneurs make their decisions, while market select enterprises more in line with the market laws according to the profit and loss system. Such a mechanism coordinates between producers and consumers in their decision-making and entrepreneurs who can produce products truly meet market needs obtain greater profits, thereby achieving stability between subjective expectations and market order in an uncertain market. Therefore, such a mechanism fails when there is failure of price signals due to government intervention, especially the decrease in interest rates caused by the expansion of the money supply, resulting in market depression.

We have discussed on the economic growth theories from Keynesian and Austrian School. The former has certain effects when facing economic crises but will result in blind expansion of monetary policy and the inefficiency of resource allocation; the latter emphasizes the nature of liberal order and the role of entrepreneurs' innovative spirits and is more in line with the nature of Adam Smith's economic research. However, no matter which theoretical thinking, they have certain defects and limitations when applying in the field of digital economy, which are mainly reflected in the following three aspects:

First, traditional theories fail to introduced non-economic factors into the path of economic growth analysis systematically. In fact, we have seen the non-economic factors impact on the entire economic growth throughout, especially in the early stages of economic growth. Therefore, we can truly know the essential causes to economic growth only by examining placing economic growth under complex economic environment and from social dynamics perspective. Therefore, the research on digital economics will dedicated to expanding the economic theories integrating the economic and non-economic factors, focusing on summarizing the internal logic between various economic challenges in the society and related decision-making behaviors.

Second, the traditional framework fails to accommodate the process of new production function generation and diffusion, that is, it is of poor scalability. However, I think we should not consider the study on economic development theory a static goal, but as a dynamic and iterative academic research process. Therefore, the framework constructed must be highly scalable. Particularly, during the development of digital economics, as new technologies such as the Internet, artificial intelligence, blockchain, and the Internet of Things emerge, there are more variables affecting production efficiency, and changing economic development factors, posing new requirement on the scalability of the economic framework. At the same time, how the complex connection between science, invention and innovation create a new production function should be analyzed, and modern economic growth and the experience accumulated in the long history of human civilization connected.

Third, the traditional economic growth theory fails to explain the connection between the business cycle and the economic growth process systematically and in depth. Economists even separate the business cycle from the economic trend in most of the time. We should break away from this static and closed economic research framework, and construct a dynamic equilibrium path or even study from an imbalance perspective. In particular, we have to see what a huge influence of technological changes on the economic world and what a dynamic S-shaped development process in economic development.

The above is the shortcomings we have discussed about traditional economic growth theory and Austrian economics' refutation to traditional economic growth theory. Based on this, we have discussed some of the work directions needed for the digital economics. We can understand characteristics of this discipline if mastering this part: On the one hand, we should inherit and explore the parts of valuable thoughts and on the other hand, we have to propose new frameworks and theories to make up for the shortcomings in the traditional economic theories to explain the new economic trends and technological development directions.

A New Perspective of Digital Economics

Based on above two parts discussed, we will explore the methodology of digital economics research in detail now. Due to space limitations, we will only discuss on two basic logics:

First, discuss the logic in the evolving of the structural paradigm of digital economics, and put forward the hypothesis that the network-based economy is a more general paradigm. We mainly introduce Lin Yifu's discussion on new structural economics to explain the internal logic of the development of growth issues in digital economics.

Second, pick up the heterogeneity issue of the evolution of economic growth, that is, the basic thinking about the impact of heterogeneous capital on the growth of the digital economy, which is a different viewpoint from those of traditional economics. We can understand the inherent laws in digital economic growth only after having mastered the elements of heterogeneous capital and the factors determining the complexity.

According to the two aspects discussed, we can basically understand why we consider digital economics a more general framework for digital economy research, as well as a framework with explanatory power and diffusion in the field of digital economy.

First of all, traditionally, the digital economy is considered a special economic development phenomenon, since most digital economy development theories fails to fit traditional economic theoretical research. However, this view is not accurate. The analytical thinking on traditional economics can be used, but we should pay attention to the parts of value such as the endogenous growth theory, and discard the thinking on general equilibrium. Since traditional economy studies the atomic world, which is interfered by non-market factors inevitably, for example, non-market behaviors due to policies between different countries. Therefore, the formed market mechanism has a logic of a not fully free market. So, we can understand the market in reality as an incomplete market network. However, the digital economy is a market with the target studied on driven by technology and innovation and there are few non-market factors as the traditional market does. That is to say, the traditional economy is a partial and incomplete network economy, while represented by the Internet, the digital economy is a more complete network economy.

According to the theory of digital economics, network covers not only the Internet, but all economic phenomena which can be understood from three perspectives:

First, economics is known as the crown of sociology (it can be understood as the essence of all sociological theories), while society itself is a complex network of different groups of people, so digital economics see human behavior and economic phenomena from the structural dimension, sharing a higher pattern than the existing traditional economics.

Second, based on the logic of the network economy, digital economics has established a wave-particle duality of quantum thinking to expand the dimension we see economics. We can see both cooperation and selfishness in economics, both independence of node and cooperation with each other which explain the dual structure theory of community organization in the digital economy.

Third, the network economy constructed by digital economics shares a logic of a organic and dynamic theory. Therefore, based on the traditional economic theory with self-interest as the core, we can explore how a cooperative network generates greater benefits. The conclusions drawn are not limited to the "selfishness achieving altruism" structure of traditional economics, but cover an explanation of economic phenomena brought about by mutual benefit and mutually beneficial communities. Therefore, according to digital economics, it is the network-based community organization that is the characteristics of the future economy. Traditional enterprises study only special cases of the network economy in the atomic age which makes traditional economic theories based on atomic phenomena an iterative paradigm and be more explanative.

Then, with reference to new structural economics thinking from Lin Yifu, we will discuss in detail on the structure of the market. The new structural economics bears a main connotation of believing that the factor endowment and structure of each economy at each time point should be the focus of economic research, since each economy has fixed capital, labor, and natural resources at each time point but they are constantly changing over time. Based on this, the new structural economics proposed a very important core point of view: each economy bears endogenous industrial and technological structures, dependent on factor empowerment at this time point. This conclusion is one of the important theoretical foundations for us to analyze the development of the digital economy in digital economics, allowing us to know that the factor endowments brought about by the network economic structure determine the basic logic in the development of the digital economy as well as development laws of most business organizations. Historically, the growth of the Industrial Revolution does not lie demographic factor, but the improvement in the productivity and quality of each worker due to the innovation of industrial technology. Besides, as the technology and industry updates, the production scale and market exchange range expand, allocating different economic factor resources into different industries. As exchange expands, strangers collaborate with each other in larger scale, during which a modern company system and corresponding laws were given birth to accompanying the emergence of capital and the constant improvement of financial system. This is the inner logic of the modern economic growth theory, that is, as the economic foundation constantly changes and the productivity improves, the infrastructure and institutional arrangements are constantly changing to promote the reduction of transaction costs. The biggest change does not lie in the basic labor factors, but that the changes in the internal structure promoted transaction costs to be reduced.

From this perspective, we will explain the reasons to the rapid development of the digital economy. It is precisely because the digital economics mainly studies on network organization, which has a better and more complete structure, compared with the characteristics of corporate organization (the more complex the corporate organization, the higher the cost), the higher the complexity of the network organization, the lower the transaction cost it is of, that is, the network organization realizes lower transaction costs. In the traditional economics, the core is to raise efficiency by division of labor when discussing how an enterprise is possibly to survive, however, in this case, the cost for transaction is raised within the enterprise. For the online community, the more complex, the lower the transaction cost it is of, while solving the professional problem of division of labor.

In summary, we have discussed the growth theories in digital economics from three different perspectives: digital economic growth equations, the limitations of traditional growth theories, and a new perspective of digital economics. As a matter of fact, the former two is about the understanding of the framework of traditional economic growth theory and the analysis of its shortcomings, while the last one is to explore the growth theory of digital economics from the perspective of new structural economics. The point is that in digital economics, people mainly look at all subjects from the perspective of network structure, believing that the traditional economy is an incomplete network while the consensus-based community organization is a complete one. Actually, the network provides a flat and low-friction (if not friction-free) method to allocate resources. It is flat, so there is no high transaction costs like the market, and it is of low friction, so there is no internal transaction costs brought about by corporate stratification mentioned by Kos earlier. On the one hand, the digital economy is a network structure-based infrastructure, and on the other hand, corporate organizations are also becoming networked, reducing transaction costs further and improving production efficiency. This is how we see the growth theory of digital economics form a basic point of view, and how we establish the basic framework for study on digital economic growth.

Chapter 4
Complexity Economic and Order of Evolution

In this chapter, we will have a discussion on the perspective of complexity economic and the economic order based on evolution. In fact, as early as Alfred Marshall, the British economist wrote the "Principles of Economics" [15] (an economic classic considered as great as Adam Smith's "The Wealth of Nations" [16] and David Ricardo's "On the Principles of Political Economy and Taxation" in history) and became one of the founders of neoclassical economics, as we see, the mainstream economics had been exploring following Marshall's footsteps. Marshall's view about demand and supply, marginal benefits, and production costs, etc. has been indispensable in economics courses for university students. However, as the times develop, especially as the digital economy advances, this economic framework based on the continuity assumption has increasingly failed to explain technology, innovation, and the evolutionary economic order discussed in behavioral economics. For more than 200 years, the traditional economics had had its main models and frameworks from the ideas of classical mechanics in physics, while evolutionary economics has tried to studied on the diversity and adaptability in economics as the objects, and to analyzed based on ideas in biology. This is the logic we will focus on in this chapter. To understand the basic ideas in the development of the digital economy by constructing a framework based on evolutionary economics.

In other words, to understand the basic logic in the digital economy, we must establish an evolution-based economic framework which is the focus of our work, while complexity economic and evolutionary economics are the basic logic when we study on this framework.

In order to study this part of the content, we will divide it into three parts for discussion: "Principles of Economics" and continuity assumptions, emergence and evolution-based complexity thinking, complexity economic and the principle of increasing returns.

Z. Liu, *Principles of Digital Economics*, Contributions to Economics,
https://doi.org/10.1007/978-981-16-9020-4_4

"Principles of Economics" and the Continuity Hypothesis

First, let's have a discussion on the theory of Alfred Marshall, the father of neoclassical microeconomics (he is one of the candidates for this title, and the other is Leon Walras, who proposed the general equilibrium theory), especially his discussion on the continuity hypothesis in the work "Principles of Economics". Due to space limitations, let's have a look at three basic points here:

First, economists should pay attention to and define the scope of their research on economics, rather than choose an extensive definition or field of economics for study.

Second, he believes that it is the primary task of economics to eliminate poverty, so his research should be carried out around the causes and solutions to poverty.

Third, Marshall realized that economic research is very complicated, but he adopted the partial equilibrium analysis method to break down complex issues, that is, to isolate a part of the economy to be analyzed for research, and get a closest result. Such a method established the mathematical thinking and hypothetical model in neoclassical economics.

Then, let's discuss a basic way of thinking involved in his book "Principles of Economics", which is also a basic thinking for human to cognize the world: the continuity hypothesis. This is the concept that we should focus on and the reason why we should rebuild an thinking to understand economics. The general point of view is as follows:

First, Marshall believed that the evolution of economics is a continuous process. He said, "Economics is—and must be—a slow and evolving science... The new theories supplement the old ones," expand and develop, and sometimes revise the old doctrines. "In addition, due to different focuses, the old doctrines are often given new explanations, but rarely overthrown." This view allowed Marshall to iterate and demonstrate when he studied based on the views of his predecessors. For example, due to the influence from Cournot and von Tunnen, two mathematical economists in the early stage, he transformed the economics of Ricardo and John Stuart Muller into mathematics for research.

Second, Marshall believed that human time and human language are continuous, "The time factor is absolutely continuous itself: Nature does not divide time absolutely into long-term and short-term; but due to the difference in unconscious degrees, the two are mutually integrated". The continuity of human language depends on the continuity of human experience. "The principle of continuity can also be applied to the use of nouns." Marshall connected this continuity thinking to the way humans get to know the world.

Third, Marshall believed that social thought and system evolution are continuous without sudden changes. "The concept of continuity of development is common to all schools of modern economic thoughts." Naturally, this kind of thinking allowed him to adopt mathematical continuity assumptions for study on economic principles, since without the aid of mathematical symbols, it would be very difficult to understand economic principles based on continuity in a complete and intuitive manner.

Based on continuity assumptions above, Marshall discussed the impact of demand on prices (the concept of price elasticity of demand), the concept of consumer surplus, the consequences of taxation and welfare and other important microeconomics principles. Borrowing from ideas from classical economists, he constructed a series of models and charts for understanding of economics based on mathematical methods and continuous thinking.

It should be noted that Marshall specifically pointed out that these concepts are constructed abstractly and theoretically. This theoretical progress was made when there was no such reality. Marshall himself also claimed that prices depend on the effects of a complex system in the long term. Therefore, we can deduce that the theoretical model of neoclassical economics is a kind of compromise, and the consequence of which has been magnified in the process of economic research that is increasingly modeled. In the next, we will discuss if the framework of neoclassical economics fails to explain the world, then what kind of framework should we adopt to understand economics? The answer is: The economic thinking based on emergence and evolution.

Complexity Thoughts Based on Emergence and Evolution

After we have discussed the continuity hypothesis of Marshall, we will explore the theories of evolutionary economics and complex economics in this chapter. First, we will have a look at the theoretical ideas of Western evolutionary economics which we start from two perspectives: the natural scientific foundation of evolutionary economics and the application methods of evolutionary economics.

The period of when the Western economic theories took shape was also the time when the classical physics in the natural sciences developed, and when Newtonian mechanics flourished. In the "Philosophiae Naturalis Principia Mathematica" [17] written by Newton in the eighteenth century, the research results of Galileo, Descartes, Kepler, Huygens, Hook, etc., were concluded, not only triggering the ideological revolution in natural sciences, but also laying the foundation for people world view and cognitive view nowadays. It was the classical mechanics that influenced and produced the classical economics, so the classical economics has very obvious limitations concerning social space-time basis and logic basis. The social space-time basis limitation refers to the fact that mainstream Western economics regards human society as a relatively static process of economic development, and pays less attention to changes in economic development due to the evolution of production relations. The logic basis limitation is the market price theory derived from the utility function and the demand curve in mainstream Western economics. In fact, this static model obviously fails to explain many economic problems in the real world.

These defects have existed, bringing our attention to the thoughts of Western evolutionary economics, that is, biological theories. Here we will discuss three basic theories in biology:

1. Comprehensive theory of evolution, that is, the integration of Darwin's "natural selection theory" and Mendel and Morgan's "population genetics", and then the integrated theory of "biological evolution" was completed by combining with molecular genetics. At this stage, Darwin's natural selection theory explained the process of biological evolution in a systematical manner.
2. Universal Darwinism. This is a theoretical idea put forward by Richard Dawkins, a famous biologist, who believed that the core concepts of variation, duplication, and selection of Darwin are applicable not only to biological world, but also to other systems that have been open and constantly evolving, such as the evolution of human civilization and society, etc. "Universal Darwinism" study based on biological thinking by combining the evolutionary phenomena of different disciplines to understand how different systems to develop after consideration of the evolution of complex systems.
3. Lamarckism. Although Lamarck's theory of homotropic inheritance has been falsified in biology, it has received attention in economic research. Lamarck's theory helps economists in their understanding of the ecological system of human society based on the idea of initiative to variation. Because human society is subjectively active, this theory can be explained to a certain degree in economics.

We have introduced the thoughts and applications of evolutionary economics above. The evolutionary economics develops, giving birth to a series of studies on new economic theories and thoughts. Theories from institutional economics, Austrian economics and complexity economics have applied and studied the evolutionary economic in a systematical manner. Here we will focus on the thought of complex economics, while introducing the analysis of evolutionary economy in other fields.

Next, we discuss complexity economics. As Klein said, "The Keynesian school has never considered deriving a theory based on individual society and commodity groups from the basic theory of individual and single commodity." But for economics, it is necessary to realize a theoretical thinking unifying microeconomics macroeconomics, while complexity economics provides an economic thinking system with emergence and evolution as the basis.

First of all, we need to study on economic thinking in complexity economics. The neoclassical micro-economy is centered on the price mechanism, while essentially market prices are determined by social collective choices, which are closely related to the differences in preferences of micro-individuals. Therefore, it is impossible to derive the macroscopic laws in a natural manner through adding up the logic of microscopic individuals. From the perspective of complexity economics, economy is treated as a complex system and the theoretical basis is to take the transactions of different individuals in the system which is the logic thinking in economics from a dynamic and mutually systematic viewpoint. Since in a non-linear system far from equilibrium, the chaotic and disordered microscopic individual behaviors will show macro-scale synergy, coherence or other orderly order, which is based on simple systems constantly evolving and emerging. Therefore, the microeconomics can be connected to the macroeconomics by simplifying and structuring the complexity system. Complexity economics studies on the laws of the system while paying little attention to the individual composition in the system, so what we should do is to describe which key variables have constituted the nonlinear equations of these macroeconomic system, rather than discuss the behavior of economic individuals in detail. Therefore, complexity economics has been based on the macro-system theories to construct an economic thinking to discuss key elements and variables while introducing information theory, sociology, and innovative technology-related theory.

Next we will sort out three basic concepts: complexity, evolution and emergence.

First, we will discuss complexity. In fact, this concept generated in the research process of computer science; especially, in artificial intelligence algorithms, the idea of complexity in the research on neural network algorithms is a key concept. To sort out this concept, the author recommend a book "Complexity: A Science Born on the Edge of Order and Chaos" to you. It describes the history of how American scholars construct around complexity research, which is called A New Enlightenment Movement.

Here, we will give a brief description of the concept of a complex system. It is a network composed of a large number of elements. In the network, there is no central control and complex collective behavior and complex information processing generated based on simple operating rules, and adaptability is form after learning and evolution. The complexity is the non-linear form between the chaotic part and the whole, due to which, we fail to understand the whole through the part, that is, it is not possible to access to the nature of the network as a whole based on restoring methods or thoughts. The so-called complexity science is to study how the interacting elements in such a network system generate an overall pattern, and how the overall pattern causes these elements to change. When we have understood these basic concepts, combined with the basic organizational structure of digital economics we discussed earlier—a consensus community organization constructed from a network perspective, we understand why it is not possible to understand digital economics before we understand the complexity.

To discuss the concept of evolution, we have to return to the discussion of Hayek, an economist. In his last book, "The Fatal Conceit" [18], there are several well-known appendix serving as a foundational for contemporary evolutionary theory. According to Hayek's argument, complex phenomena occur after simple phenomena constantly evolving. The micro-level subject fails to foresee the characteristics after the complex system is formed, and the key to our understanding of system phenomena is the characteristics embodied by these macro-orders. Hayek also pointed out there are three most basic laws of order in the evolution of human society: customary law, natural language and currency, and he gave a conjecture on currency similar to anarchism. We will discuss this topic in our subsequent discussion on the concept of digital currency, especially on the digital economy order, that is, how the basic order evolves in a natural manner.

Finally, let's talk about the concept of emergence. The so-called emergence refers that when individuals in the system follow simple rules and integrate through local interactions, some new attributes or laws will occur at the system level all of a sudden. This concept is corresponding to the above-mentioned evolved system having some characteristics that simple elements do not possess. Complex phenomena generate as a result of the emergence.

According to Hayek's view, language, human social networks, and the interaction of neurons in human brain are complex networks emerged. So why should we study on complexity phenomena? Just because of the uncertainty; the real world we face is uncertain, rather than the ideal model constructed in traditional economics to satisfy a lot of assumptions. Therefore, we need to adopt an academic framework fitting the real world to understand the world. Frank H. Knight, one of the most influential economists and thinkers in the twentieth century and the founder of the Chicago School, raised that uncertainty is inevitable. As long there are enough interacting actors, simulation researchers have always failed to foresee the macro order of micro behavior no matter how simple the rules of behavior. Therefore, the order of economics should be based on hypothesis of rational incompleteness, rather than the so-called "rational man" hypothesis.

In summary, we have discussed three basic concepts in complexity thinking based on emergence and evolution: complexity, evolution, and emergence to understand that the real world takes shape based on the rules of evolution and emergence, and we are able to understand the uncertainty of the external order only by use of complexity thinking to understand the external world rather than the thinking of the hypothetical model.

We also explored the difference between the methodologies adopted in the researches of complex economics and neo-classical economics for economic studies. Based on this point of view, Brian Arthur, whom we mentioned above, studied complexity economics and the principle of increasing returns, studying economics from a systematic perspective of complexity, which is the most important method in the research process of digital economics.

Complexity Economics and the Principle of Increasing Returns

After we know the theories of complex economics, let's discuss on some basic ideas and theories of complex economics, especially the thinking about equilibrium theory and increasing returns in complex economics.

First, we will have a look at the frontier concepts of evolutionary economics. Due to space limitations, only three important concepts are introduced here: organization and self-organization, evolutionary game theory and co-evolution theory.

Organization and self-organization are the structural forms most common in biology, and it provides enlightenment to the theory of economic development when the theory of organization and self-organization are applied to economies and economic behaviors in evolutionary economics. According to evolution theory, the organization is a means. For example, Dawkins, a biologist considered genes as the "replicators" and organisms as "vehicles" of genes; Hull, a biologist regarded genes as "evolver", and organisms as genetic "influencers". According to the theory of evolutionary economics, in economic organization process, information is evolving rather than copying, and economic organization is not merely a means of delivery, so Hull's theory is more practical. In addition, according to this inference, Hayek emphasized the distinction between spontaneous order and artificial order in the economy which is meaningless, since all artificial order can be regarded as the evolution of genetic spontaneous order, so the economic system is generated when in complete spontaneous evolution. In fact, we believe that the economic and biological organizations have no differences in other characteristics other than the difference in the ability to process information.

Therefore, a new economic thinking is required to be constructed based on evolutionary thinking, that is, thinking based on complexity economics. Let's give an introduction to the basic ideas of complex economics created by Brian Arthur:

1. Different from views of neoclassical economics, he believed that disequilibrium is a natural state, and was endogenous as early as the economy took shape. This view is completely different from that of neoclassical economics. For example, Samuelson (the writer of the most widely used version of "Economics" of university textbooks) believed that even if really exists, the disequilibrium states is temporary and non-persistent.
2. There are two reasons for disequilibrium to exist. One is the existence of uncertainty, that is, the choice in economics is actually made under an uncertain environment, rather than a rational choice from complete information. There is no sufficient information, so the optimal choice does not exist. Another reason is that due to technological innovation and technological change, the economy has been developing in an unbalanced state. As Schumpeter said, "There is a force in the economic system that can break down any possible equilibrium" (We will elaborate on Schumpeter's research on technology and innovation theories later).
3. The disequilibrium phenomenon in complex economics are able to self-reinforce and feedback positively, which means that the complexity economics is studied on based on the application of complexity series of theories, while complexity systems is characterized as positive feedback, or to be accurate, positive feedback and negative feedback exist and interact with each other at the same time. If there is only negative feedback, the system will quickly converge to an equilibrium state, becoming a static "dead system". When there is only positive feedback, the system will deviate from equilibrium, coming into an "explosive" state. The two feedback effects at the same time, resulting in a complexity system. Different economic structures form and disappear, constantly evolving and iterating in this process, thus forming a "living system." This recalls us to what we said before to look at the evolution of economics with a structural approach, which is also an important reason why digital economics chooses the thinking and framework of complexity economic theories for research.

Due to the characteristics of the complexity economics mentioned above, we know that complexity economics constructs an economic theory of "increasing returns". Due to space limitations, we will only give a simple explanation. The so-called increasing returns mean that a technology is in a leading position to get a more leading position due to the action of the "positive penalty" mechanism. Moreover, increasing returns is a random phenomenon, determined by the natural selection process in the evolution paths of different technologies.

Complexity economics believes that economic evolution is a process of natural selection realized under a certain set of small random events. Therefore, two basic elements should be focus on in the study of complexity economics: first, the nature and evolution of technology, that is, the choice of technology determines the path and efficiency of economic evolution. Second, there is inertia in the technological evolution, institutional changes and cultural elements in human society similar to that in physics, so the idea of increasing returns is path-dependent.

In summary, we have discussed three main topics: "Principles of Economics" and continuity assumptions, complexity thinking based on emergence and evolution, and the complexity economics and principle of increasing returns. We have understood the shortcomings and deficiencies of mainstream economics after analysis of academic achievements of Marshall and Walras, two most important scholars in neoclassical economics. We introduced the concepts of emergence, evolution, and complexity to establish the basic logic of complexity thinking.

Finally, we introduced the basic concepts of evolutionary economics and the economic thinking of increasing returns, especially we analyzed the concepts of self-organization, evolutionary games, and co-evolution, which correspond to the growth model of digital economics we mentioned earlier, providing a corresponding ideological basis for us to introduce the value network-related theories later. Not merely a kind of disciplinary thought, the evolutionary economics is a cognitive system different from the physical thinking of traditional economics. By the systematic discussion on evolutionary thought and complexity economics, the author hope you can have a deeper understanding of the basic methodology system established in digital economic theories.

Chapter 5
Technological Elements and Enterprise Innovation

In this chapter, we will have a discussion on the relationship between the two most important elements in digital economics: technological elements and enterprise innovation. We have discussed something about innovation above, mainly from the perspective of development process of the IT information technology revolution and the driving force of innovation. In this chapter, what we discuss are the relationship between technological elements and innovation in the era of digital economy as well as the relationship between technological development, innovation and economic growth and evolution. We can hardly understand the nature of technology and the core role of innovation as the digital economy develops, and deduce the basic theoretical basis that innovation is the most important task for entrepreneurs in digital economy before clarifying this chapter.

We divide it into three parts for discussion: Schumpeter's theory of enterprise innovation, technological evolution and economic structure, entrepreneur theory and innovative thinking.

Schumpeter's Theory of Enterprise Innovation

First, let's get to know Joseph Schumpeter, an Austrian economist. In recent years, because of China's innovation wave in the field of digital economy, especially the innovation wave in the mobile Internet, his theories and ideas have become known and recognized by many entrepreneurs.

In fact, on the one hand, Schumpeter had spent all his life advocating the liberal capitalism economic system, and fighting unremittingly with Keynesian theory, but he failed to gain a reputation corresponding to his achievements. On the other hand, he influenced masters in the field of business management such as Drucker, so Schumpeter's theory was very popular with the entrepreneur and innovator. However, in recent years, few comprehensive and in-depth discussions about Schumpeter's theory emerge, instead, there are more discussions on "disruptive

Z. Liu, *Principles of Digital Economics*, Contributions to Economics,
https://doi.org/10.1007/978-981-16-9020-4_5

innovation" which is a specious concept and the blunt application of his theory in business practice. What we are to do here is to reveal the truth of Schumpeter's economic thinking on innovation and growth theory, to help you understand the essence of this chapter.

First, let's have a look at the roots of Schumpeter's thoughts. According to him, he offered the highest evaluation and praise for his thoughts to Lyon Walras, a French economist we mentioned earlier, also one of the founders of the neoclassical economy.

In the book "Theory of Economic Development," Schumpeter made a basic assumption based on Walras' static equilibrium system theory: consider a society dominated by commerce, in which private property rights, division of labor, and free competition are generally practiced. Based on this, we know that what Schumpeter had been focus on was the innovative economic theory after the development of advanced industrial capitalism rather than capitalism in the primitive accumulation stage. On the other hand, Schumpeter offered a theory of economic development of allowing the dynamic equilibrium to occur in the system through technological incremental changes although his theory is derived based on the static equilibrium theory.

Specifically, Schumpeter's economic development theory can be studied from three perspectives:

1. Schumpeter believed that the main economic changes in the capitalist era happened in an irreversible revolutionary manner, rather than in a continuous and gradual way, that is, "the spontaneous and discontinuous transitions in the circulation channel" is at the core in the development of capitalism.

Second, Schumpeter believed that the effective response of a specific innovation subject to innovation opportunities is the key to economic growth when innovative entrepreneurs make decisions and carry out production activities in the market full of incomplete information and uncertainties which is the core in economic growth, in other words, the entrepreneur's innovation ability and the capital's recognition of the credit of the entrepreneurs are the premise to gain the market under complete market conditions.

Third, in the development, the economy iterates and changes which is depending on the discontinuous changes in the circulation channels. The iteration is never-ending, constantly replacing the original equilibrium. What destroy these equilibrium factors, include the introduction of new products and new production methods, opening up new markets, obtaining new raw materials or supply sources, and new changes in industrial organization, in short, depending on technological innovation and organizational innovation.

Based on the theoretical discussion above, we know that Schumpeter did not merely focus on static factors for his theoretical thinking, but he did it based on technological innovation and innovators as the core. Therefore, we consider Schumpeter a scholar who understood most thoroughly the two seemingly contradictory theories of economic growth and economic cycles. In the book "Theory of Economic Development", he raised several key questions about economic growth and business cycle theory, offering very thorough and creative thinking about these questions.

1. Why is economic development is advancing in a cyclical manner rather than in a constant way? The reason lies in innovation, in other words, the combination of the original technological elements and production factors which is not "evenly distributed over the time...but in a discontinuous swarm", to put it simply, it emerged through evolution and emergence as we discussed before.

Second, why is there emerge in the economic cycle phenomenon? Why does this emergence spread? Schumpeter believed that due to the emergence of a small number of innovative entrepreneurs, a phenomenon of diffusion will occur. Some entrepreneurs succeed in innovation, attracting other innovators to join in, resulting in be large-scale innovations and industrial clusters that we often see in the Internet or other fields of the digital economy. In short, the combination of new technological elements and innovative elements is the fundamental feature of economic prosperity, explaining the internal logic of economic prosperity spreading to the entire system.

Third, how we should understand the surplus value of the market, that is, the interest theory based on the premise of the innovation theory? Schumpeter believed that, different from the theory of interest in general static cyclic systems, more profits in the network constructed by innovators occur when the total revenue brought by the positive feedback due to the innovation on all aspects of the system exceeds the total cost. In other words, the "creative destruction" activities brought about by innovation replace unprofitable economic activities, thereby creating profits, while the innovation diffuses, making the original innovation unprofitable, so entrepreneurs have to innovate further. The process of investment and wealth expansion is to reinvest the profits of such endless innovations.

In summary, we discussed the theory of innovation and economic growth of Schumpeter who had been focused on the cyclical issues in the process of economic development in the capitalist era with specific elements including the static circulation equilibrium model, the internal and external driving forces of economic change, innovation and the role of entrepreneurs, etc. What we are concerned about is his assumption of a sufficiently developed free market and the innovation-related cycle theory, since actually, the digital economy discusses a market fitting Schumpeter's assumptions more than the traditional industrial economy or capitalist economy, so his theory is more explanatory. Schumpeter's theory had not received adequate attention in the industrial economy era since actually his hypothesis is more suitable for the scenario of digital economy development, and our subsequent discussion on the economic development model based on network organization is completely in line with Schumpeter's theory of technological innovation and economic growth.

Technological Evolution Theory and Economic Structure

In this chapter, we will discuss the relationship between the impact of technology and complexity economics in detail. On the one hand, the destructive impact of technology corresponds to Schumpeter's theory of technological innovation. On the other hand, seeing the iteration of economic structure from the perspective of complexity economics will contribute to better understanding of the relationship between technology and economy. Here we discuss it from three perspectives: first we discuss the technological evolution and the evolution of economic structure, then several conclusions brought about by the emergence of complexity economics, and finally the relationship between technological evolution and economic evolution.

First, let's have a look at the logic of technological evolution and the evolution of economic structure. In fact, in traditional economic theories, technology is equated to the production function, and economy is considered the "containers" of these technologies. When new technologies are introduced, the function changes as well as the total production volume and the production efficiency, releasing other related resources (labor, capital, natural resources, etc.), allowing more capital to be invested in research and development of new technologies. Then the economy shifts from one equilibrium to another and achieves endogenous growth. This model makes technology the background, while emphasizing price and quantity factors; in addition, it fails to clearly explain the impact of the generation and evolution of technology on the economy.

Here, we are to establish a simple model for economic growth, including the agricultural sector, industrial sector and technical production sector which replace the main economic forms before the industrial revolution, after the industrial revolution, and the digital economy. In addition, from the perspective of the long-term economic development process, in most countries, the economy develops from an agricultural economy to an industrial economy, and then to a process of endogenous growth through technology. Here we focus on the role of technology in the endogenous growth of the digital economy.

Based on this model, we can derive the following basic conclusions. To realize the endogenous growth of the digital economy, that is, growth driven by technology, it should meet three basic conditions:

First, there must be compatible coordination between different sectors, that is, the endogenous growth of technology should not only rely on the technological production sector, but must be realized based on development in the industrial and agricultural sectors. This is what I have always emphasized that, the digital economy can not be independent from the real economy for development, and I do not agree with the view that the digital economy is the only destination for human economic development in the future.

Second, the economy can grow in a dynamic and sustainable manner only when the increasing return to scale of the knowledge sector brought about by technological growth exceeds the diminishing return to scale of the commodity production sector. Due to the productivity development and the replacement in employment brought by the technology sector, the value it creates is higher than the cost it bears in the development process, which is the basis for the digital economy to develop. Third, there are multiple directions and multiple paths for different types of endogenous growth of the digital economy, among which the ideal one is the sustainable and stable growth under dynamic equilibrium, when different variables are transferred in a balanced manner in roughly the same proportions. Achieving balanced, dynamic and stable growth remains an important goal whether it is the transformation of the agricultural economy into an industrial economy or the development of an industrial economy into a digital economy.

We should note two points: First, the digital economy develops in a time when demographic dividends in the industrial economy era drop and the global financial capitalism proliferates. It is a more complex and unstable environment, and a risk and challenge the digital economy has to face. Second, as the digital economy develops, it solves problems through the dual effects from the industrial economy and the digital economy, rather than relying on one of them. Take China for an example, although it has become one of the two countries with the best digital economy development, it relies on the great right of speech gained by the Made in China industrial system in the global division of labor. Broadly speaking, whether the industrial economy or the digital economy, they have gained the driving force for rapid growth under the logic of the knowledge economy, in other words, technology is the most important driving force for endogenous growth.

According to the views of complexity economics, we regard technology a broad concept, which is defined as various tools and means used by humans to achieve their goals, including production technology, organizational structure, laws and systems, etc. Therefore, we include human knowledge and innovation into the concept of "technology", while new technology is the innovation obtained after continuous combination and iteration of the original technology. In short, we consider technology a constantly evolving and iterative system: new technical elements are constantly generating from existing elements, while being combined into more complex elements to play their roles in the ways of evolution and emergence based on different scenarios.

At the same time, from the perspective of classical economists, we define the economy a series of arrangements and activities based on which the society meets the needs of each individual. Then the economy is a process with the starting point of its production tools and it emerges based on its own arrangements and technology. From this perspective, the economy is a ecosystem with emerging technologies, and technological elements are constantly iterating in the economic system to promote the economic structure to change constantly.

We should note that new technologies are usually introduced in the form of technology clusters, forming the phenomenon of scientific paradigm shift that Kuhn said. For example, the first scientific and technological revolution was centered on steam engine technology, while the second on electricity, and the third on computers. Therefore, new technological elements are constantly emerging which promote the scientific paradigms to transfer through the phenomenon of clusters and create new human economic activities when combined with existing social and commercial elements, inspiring humans in the creativity and imagination and forming new business models and new processes. A small number of enterprises develop into monopolies, while most enterprises promote the economy and society to change based on new technological innovations. This process shows the relationship between technological evolution and economic structure evolution.

After having known the characteristics of complexity economics, we can deduce its three basic effects on economic theoretical research naturally: cognitive foundation, structural foundation, process and perspective based on process and emergence. Due to space limitations, we will have a brief discussion on the relevant content.

The so-called change in the cognitive foundation refers to that in the cognitive paradigm of neoclassical economics, the economic behavior agent is defined based on the assumption of the so-called "rational man". In other words, as an agent, each individual evaluates the uncertainties in accordance with the probabilistic thinking of absolute rationality before they update the information according to the Bayesian method to adjust the assessment, and choose the action plan maximizing their utility expectations. In fact, this process looks more like the way of "artificial intelligence" making decisions, and the theory of behavioral economics has demonstrated that human behavior is influenced by a lot of irrational factors. Therefore, from the perspective of complexity system thinking, we do not think there is a single, dominant cognitive model. Human behavior is a decision made based on a complex thinking process. The so-called optimization in the standard sense is an assumption for simulation and optimization in the real world. Because of this, to study the economics of the real world, we should try to use a more complex, specific and uncertain cognitive methodology, so as to understand complex systems and economics based on emerging orders.

The so-called change in the structural basis happens because traditional economic theories are based on the full game between people, leading to a simple and extreme interaction structure: either one person faces everyone, or everyone faces everyone. This structure is similar to the "all-to-all war" mentioned by Hobbes in "Leviathan" [19] and it is a very abstract and simplified game model, ignoring the mutual assistance elements in human behavior and complex structure. In fact, based on the perspective of complexity economics, what we should establish is a view that is more dynamic and networked, which has three implications:

First, the network structure is not a linear structure and it is the most important method for connection in the economy; all economic behaviors are realized in the network. We will discuss later that in digital economics, the main body is the network rather than the enterprise since one of the theoretical foundations is that the market is an incomplete network, while community business ecology based on consensus in the digital economy era forms a network that is more complete.

Second, economic behavior is the social order developed through emergence, and these orders have been structurized through institutions. This theory is actually the economic behavior that people constrain their economic behaviors through specific contracts. In the digital economy era, these contracts are placed on the network through technical means (that is, smart contracts) which is an important starting point for us to understand the economy in the digital economy era.

Third, economic entity is of a recursive structure, and the organization generates organizational levels that are more complex by continuous iteration and recursion. This theory serves as a basis when we discuss how the complexity of the organization forms and the theory of multiple structures. We can hardly understand the main body of the digital economy era and the iterative and evolutionary methods before we understand this structure.

Finally, we will discuss the relationship between technology and economic growth. Here, I would like to share some theoretical thinking of digital economics from the perspective of new structural economics. Complexity economics explains the relationship between economic growth and technology, while new structuralism brings us a new perspective and point of view when we try to understand the logic of endogenous growth of technology:

First, the economy emerges from technology. Therefore, not only will the economy re-adjust its structure and elements as the technology evolves, but also will it reconstruct and iterate due to the evolution of technology. In short, the economy has the characteristics changing with changes in technology. This is the reason why we discuss the subject of digital economics specifically; economics also needs to adapt accordingly to the iteration and changes in technologies.

Second, technology builds the structure of the economy, while the economy regulating the creation of new technologies, and therefore also regulating the creation of the economy itself. From the perspective of a long economic cycle (decades), the economy has been a process of continuous formations and interactions and crashes. And from the perspective of a longer cycle (thousands of years), the economy is a process of continuous creation and growth driven by technology actually. We had discussed Lin Yifu's New Structural Economics, which takes the factor endowments at each point in time as the entry point for economy studies. Differences in the endogenous structures of technology and industry of each economy at each point in time bring about differences in endowments, leading to the differences in economic structure.

Third, technology will cause structure and institutional arrangements to change, leading to recombination and changes in others factors except the economic factors, such as laws, culture, and the basic order of society. Once technology changes, the economy will combine or be arranged in a new manner, around which other factors will also change.

I have raised the concept for future enterprise to evolve from a traditional business organization to a "niche economy"; for enterprises, they should also develop in the way of "governing the country". The nation develops the economy in order to raise the economic growth, and the intermediate goal is to improve the industrial structure, that is, to achieve economic growth through being improved to the factor endowment structure. In this process, it is necessary to improve the capital owned by each labor force, improve the infrastructure to increase efficiency, and improve the institutional arrangements to reduce transaction costs. The same is true for enterprises. It is necessary to increase the capital owned by participants in the enterprise ecosystem, to construct the core platform of the entire ecosystem, and to reduce internal credit costs and improve synergy efficiency by introducing new incentive mechanisms and institutional arrangements.

The above are the new cognitive revolution and cognitive perspective we introduce based on the perspectives of complexity economics and new structural economics; they are the basic logic for us to follow in the subsequent studies on specific issues in the digital economy. We hardly have adequate insight into the basic pattern and basic theories of digital economics before we master the knowledge of the cognitive and structural foundations.

Entrepreneur Theory and Innovation Thoughts

In the last chapter, we will talk about entrepreneurs. Actually, one of the signs indicating the maturity the modern economy is to transform the main body of the market from capitalists to entrepreneurs. In most cases, capitalists promote the social capital to redistribute through investment, while entrepreneurs drive the technologies to develop and the market to grow through innovation.

Therefore, to know the definition and function of entrepreneurs, as well as the differences in the ways that different schools of economics study entrepreneurs is very important for us to understand entrepreneurial innovation in the theory of digital economics as well as for entrepreneurs to understand the border of the enterprise organization.

First, we discuss how the concept of entrepreneur originated and developed. The term originated as early as the Middle Ages; it referred to individuals who can accomplish a certain goal actively. As time changed, entrepreneurs mainly referred to some risk-takers who accumulate resources by contributing their lives and wealth. After the seventeenth century, entrepreneurs mainly referred to people who took risks or provided certain services. For example, priests or architects who participated in the construction of churches were called entrepreneurs. Since large-scale projects required individual leadership and management capabilities, professional level and leadership were considered the core competence of entrepreneurs, and thus forming the theory of modern entrepreneurs gradually. In short, as early as the sixteenth century, when the term entrepreneur came into being, it had been given the identity of an adventurer. In the centuries since then, entrepreneurs had been given the identities of operator, capitalist, manager, and innovator, indicating the changing role and functions of entrepreneurs in the long history of changing times, developing productivity, and institutional changes. Correspondingly, the capitalist is defined to the individual who is able to pay in advance the wages of people involved in the production process, gaining the rights to capital goods and final products. At a certain stage, the capitalists and entrepreneurs have their identities highly coincided; but they have a big difference between them nowadays.

Then we will discuss the difference in the theoretical research methods when study from the management and economics perspectives, as well as the selected dimensions of digital economics. In the 200 years since 1830s when Richard

Cantillon [20] raised the concept of entrepreneur, from Say to Marshall, three stages of early development of entrepreneur theory have been experienced. Schumpeter defined the category and function of professional entrepreneurs, laying a theoretical prerequisite for incorporating entrepreneur theory into modern enterprise theory. The latest modern enterprise theory develops and the contract theory and incentive theory advance, almost all with entrepreneur theory as the starting point. Entrepreneurs' incentive and restraint issues are closely related to the nature and type of enterprises, corporate governance structure, corporate capital structure, etc., and modern entrepreneur theory has taken a large share in modern enterprise theory.

Whether in economics or management, the study on entrepreneurial talents is a key subject derived in recent decades, in other words, entrepreneurial theory is a interdisciplinary theory that is relatively new for research. In management, more emphasis is placed on the study of the organizational ability and leadership ability when individuals make the decisions, that is, focusing on the analysis of individuals without paying attention to the relationship between the entrepreneur and the enterprise. In the process of traditional economics research, entrepreneurs are more often considered the object of market analysis, and little attention has been given to the talents and individual initiative of entrepreneurs. However, the theory of digital economics chooses the connection between the entrepreneurs and enterprises as the dimension for core research object, the source of which are the important economists we will discuss subsequently.

Here we mainly introduce the views from three economists: Schumpeter, Kerzner, and Knight-Mises's research frameworks. They three represent three research frameworks respectively.

In Schumpeter's research framework, it determines whether a person is an entrepreneur through the dimension of innovation. He found that many enterprises fail to get a systematic written strategy, but foresee the future development of the enterprise relying on individual entrepreneur's qualities such as intuition, judgment, wisdom, experience and insight, and restrict the development of the enterprise through his values, power and will. He defined innovation as a variety of combinations, including the introduction of new products, new production methods, the discovery of new markets and demands, etc. while entrepreneurs enabling the economy to break through the original equilibrium to transit through such innovative combination activities, that is, the so-called disruptive innovation. Schumpeter believed that the important function of entrepreneurs is to achieve economic transition through innovation. As an admirer of general equilibrium beliefs of Walras, he believed that the function of entrepreneurs is to promote the economy from equilibrium to non-equilibrium, while other mainstream economists often think that entrepreneurs are pushing the economy toward equilibrium.

Then we will discuss the theory of the enterprise research framework of Israel Kerzner. As the head of the contemporary Austrian school of economics, in the book "Competition and Entrepreneurship" [21] he defined entrepreneurs as: "People who are beyond the reach of general population to keenly discover profit opportunities in the market, and have insights." In other words, he believed that the entrepreneur has a most fundamental function of discovering profit opportunities that beyond others' reach. Therefore, entrepreneurs activities in Kerzner's sense are mainly realized in an imbalanced sense, emphasizing vigilance rather than innovation. Similar to the view of Schumpeter, Kerzner also believed that entrepreneurs do not need to own capital, because they can create new markets and demands by vigilantly discovering opportunities. In other words, his entrepreneurial behavior does not happen in a complete disequilibrium system, but in an equilibrium theoretical framework, so the role of capital is no longer considered. In short, Kerzner provided a framework for original entrepreneurial research based on vigilance.

Finally, we will discuss that the theoretical framework for entrepreneurial research Knight-Mises was concerned about (this framework was later developed and improved by Peter Klein and Nicolas Foss) is a framework based on the ability to judge heterogeneous resources. They believed that since entrepreneurs need to make decisions about future uncertainties, and it is the corporate organization that does the most important subject of this decision, so the entrepreneur is the most important part of corporate theory. Therefore, we should focus on all the decisions in uncertain markets made by the entrepreneur, including the ability to create new corporate organizations and the ability to operate existing organizations, rather than merely emphasize the entrepreneur's innovation, vigilance, and talents one-sidedly. In this framework, the discussion on equilibrium theory is not so straightforward. We should note that since capital is involved when the entrepreneur makes the decisions after making judgement about the future, the capitalist and the entrepreneur are one in this theory. The entrepreneur has a core task to formulate production plans and allocate factors based on the prices of various factors to seek profits and opportunities in a market of uncertainties.

We have divided the technological elements and economic innovation in this chapter into three parts for discussion: Schumpeter's theory of enterprise innovation, technological evolution theory and economic structure, entrepreneur theory and innovation thought. The key is to understand the influence of complexity economics and new structuralism on the theory of digital economics, especially the relationship between technology and economic development. What we have always been exploring are the unstable external environment and the logic of endogenous growth faced in the development of the digital economy from two different perspectives.

Part II
Value Network Theory

Chapter 6
Value Theory of Digital Economic

In this chapter, we will discuss the evolution of the idea of a value theory of economic and the value theory of digital economic. Why is this topic placed in the first chapter of this part? There are three main reasons here.

1. With the development of economics, value theory is not only one of the foundations and core of economic theory, but also a fundamental concept in the operation of the economy. Therefore the differences in the connotations of different economics need to be identified in the course of the study of value theory. The basis of the value theory of digital economics comes from two sources. On the one hand, it comes from the traditional discussion of value theory in economic theory. On the other hand, it also comes from the study of value in the qualities of the digital economic model. Therefore, this part needs to be sorted out.
2. The basic paradigm for the understanding of the world economy by different economists has been shaped by the value theory. Among these, the traditional value theory of labour and the equilibrium theory of value have been very influential. In this chapter, the topics of prices, interest rates, capital and money can only be discussed once the theory of value has been discussed. At the same time, the new capital theory of the entire digital economy can only be studied in depth on this basis.
3. In digital economic, the economic topics related to the Internet of Value are focused on. It is essentially a study of the relationship between the Internet of information and the Internet of value and discusses the changing economic ecology in the blockchain era based on value network theory, which has led to the idea of digital economics with network organisations as market actors. This is a key starting point for the understanding of digital economics as a whole.

On the basis of the above analysis, the subject will be discussed from three perspectives: traditional economic value theory, Austrian economic value theory and digital economic value theory.

Z. Liu, *Principles of Digital Economics*, Contributions to Economics,
https://doi.org/10.1007/978-981-16-9020-4_6

Traditional Economic Value Theory

The traditional theory of value in economics has been mainly explored around two basic theories, which include the labour value theory system of classical economics and the equilibrium value theory system of neoclassical economics. The predecessor of the first theory was British classical economics, which was developed on the basis of Ricardo's labour theory of value. The second theory was created by the British economist Marshall and is based on the integrated theory of the value of production costs, the theory of marginal utility and the theory of the value of supply and demand. It is also the dominant theory of value in Western economics today. These two most important theories of value that have influenced the course of modern civilization are explored, which enable the basic logic of the value theory of economic thought to be understood.

Firstly, the context of their development is compared and discussed. In the mid-seventeenth century, the capitalist system was established in England, replacing the feudal system. This was not only key to the rise of Britain, but also to the development of a market economy and the rise of industry and commerce. With the rise of Britain, classical economics and the labour theory of value were developed. The main characteristic of classical economics was its focus on the production sphere and its relative neglect of the consumption sphere. In addition, classical economics starts with costs, and labour is the most important thing in terms of costs. Therefore, the value of a commodity is usually attributed by classical economists to human social work.

As time passed, on the one hand, the development of the productive forces was continuously driven by the industrial revolution in capitalism and its need to be explained by a new paradigm of economics, while the influence of capital, prices and other elements on value could not be explained by the classical labour theory of value. On the other hand, the monopolistic phenomenon of capitalism was constantly emerging, so theories on supply and demand were increasingly explored. Against this background, Marshall synthesised the results of the marginal utility school, the classical school and the supply school in his analysis of value and developed the theory of equilibrium value. Equilibrium value theory focuses on both supply and demand, and the value of a good is considered to be determined by the market, with neither the consumer nor the producer being able to determine the price alone. This theory not only has a very good explanatory power for market phenomena, but also conforms strictly to the ideas of mathematical economic models. As a result, it has so far been in the mainstream of economics.

The intellectual origins will then be compared and discussed. The British mercantilist economist William Petty was the first to formulate the labour theory of value in classical economics. William Petty was the first to formulate the labour theory of value in classical economics. In his Treatise on Taxes and Contributions [22], "Land is the mother of wealth, and labour is the father and dynamic element of wealth." was mentioned. It was later adopted by Locke, Adam Smith and Ricardo, and a complete theory of labour value was gradually developed. Adam Smith argued that "the price or exchange value of every commodity consists of all or one of the three parts of wages, profits, and land rent". It means that wages, profits and rents are treated as the fundamental sources of value creation. But in fact, it confuses the difference between value creation and value distribution. On the other hand, according to Smith's distinction between use-value and exchange-value, Ricardo developed a more refined and systematic theory of labour value.

Ricardo proposed that the value of a commodity was determined by the value of labour, with wages and profits being secondary. Marx's labour theory of value, Sraffa's theory of value and John. Marx's labour theory of value, Sraffa's theory of value and John Mueller's theory of the value of production costs were developed on the basis of this theory. Marshall then not only integrated and adapted the classical economic theories of Smith, Ricardo, Muller and Malthus, but also synthesised the work of Menger, Jevons, Walras and others on the theory of marginal utility. At the same time, the equilibrium value theory system was proposed by him on the basis of the analytical framework of market supply and demand equilibrium. Hicks Pegu, Arrow, Debreu, Samuelson and others after Marshall also did systematic and in-depth research on value theory within the framework of neoclassical equilibrium analysis, which made the neoclassical equilibrium value theory the mainstream theory of economics today.

Finally, the two theories of value are compared in their definitions of the nature of value. Ricardo argues that the present or past relative value of a commodity is determined by the relative quantity of the various goods produced by labour, rather than by the quantity of goods obtained by workers as a result of exchanging their labour. Neoclassical equilibrium value theory, on the other hand, focuses on the relationship between people and things, which refers to the relationship between people's needs and the scarcity of goods. The system of equilibrium value theory has two aspects. On the one hand, there is the demand of people and the utility of this differentiated demand. On the other hand. Supply refers to the price that people pay to satisfy this demand. At the heart of the theory of equilibrium value is demand rather than supply. It reflects the relationship between a certain good and what society as a whole needs, and its refers to the relationship between man and nature (a topic that will also be touched upon in the final lecture). So in neoclassical equilibrium value theory, value refers to the numerical ratio of the exchange of a good with another good. At the same time, the degree of scarcity of both goods in the market is reflected by this numerical ratio.

Austrian Economic Value Theory

The subjective theory of value is another name for the Austrian economic theory of value. In this theory, value is considered its determined by man's judgement of what is useful. Variation in the quality of labour presents no difficulty in explaining value. Thus, the value of a product is not considered to be determined by the input of the labour factor. Rather, the value of the inputs of the factors of labour is considered to be determined by the products they help to produce. We will present specifically the Austrian critique of the traditional theory of value and the subjective economic theory of value proposed by the Austrians.

First, the critique of the Austrian economic doctrine of the labour theory of value is discussed. The most famous view comes from the famous Austrian economist, Bohm-Bawerk. In the first volume of his two-volume Capital and Interest [23], The History and Critique of the Theory of the Interest of Capital, he provides a systematic critique of the labour theory of value in Marx's Capital. Bohm-Bawerk argues that purely logical deduction results in a labour theory of value. But he does not provide any empirical or psychological proof. He argued that the doctrine did not apply to the vast majority of commodities and thus deduced that the theory of surplus value was not valid. This was followed by the publication of The End of Marx's System of Thought by Bohm-Bawerk. In this book, Marx's system of political economy is systematically criticized by him. He argued that, based on the labour theory of value, the price of a commodity depends on the value of the commodity. At the same time, he argued that "there is no proportional relationship between the exchange of commodities and the quantity of value condensed into them in fact".

At a word, the Austrian scholars' critique of the labour theory of value is expressed in three main ways.

Firstly, the essential elements that do not involve labour cannot be explained by the labour theory of value. Secondly, the labour theory of value cannot explain time preference, whose value is also that of interest. Thirdly, the value of labour itself cannot be analyzed by the labour theory of value.

If the core of the value theory of classical economics is considered to be the labour theory of value, its core is considered by Austrian scholars to be the subjective theory of value. So Austrian scholars have always disapproved of the labour theory of value, and its caused Marx's theory of value and surplus value to be devalued by Western scholars. This reflects the influence of the Austrian scholars' critique of the labour theory of value.

The critique of the Austrian doctrine of equilibrium value theory is then discussed. We illustrate this with a discussion between Bohm-Bawerk and Clark on the theory of capital. Clark, according to the equilibrium theory of value, considers capital as a permanent fund which automatically generates profits in the form of interest and the phenomenon and model of time preference is not taken into account. This idea is based on the application of Walras' general equilibrium idea to the theory of capital. However, Austrian economists believe that the main flaw of the Walras model is that it ignores the fact that all parameters and variables in the model arise gradually as the behaviour of the economic system changes, rather than simultaneously. This leads to the neglect of the time element, which describes the role of an imaginary simultaneous developed variable. This simplified logic is very common in the equilibrium value theory of neoclassical economics. It also prevents the real economic problem from being solved. Bohm-Bawerk suggested that the theory based on productivity is simply a fallacy of the idea of objective value. He based his theory of equilibrium value on the idea that value is determined by the historical costs incurred in the production of a product service. In fact, however, costs are determined by prices. The origin and existence of interest must be independent of capital goods and dependent on the subjective time preferences of man.

Finally, the subjective theory of value of the Austrian school is discussed, starting with Bohm-Bawerk. Bohm-Bawerk argued that every process takes time, and that before the goal is achieved it necessarily passes through a series of stages embodying highly heterogeneous and differently combined forms of capital goods. The price of capital goods is therefore determined not by the cost of production, but by their discounted value to future productivity based on the interest rate. Simply put, neoclassical economists believe that theory is determined by both the supply and demand for capital, with supply determined by the subjective perception of time preference and demand determined by the marginal productivity of the entrepreneur based on capital. However, Austrian economists see entrepreneurs as the end of workers and factors of production when they need capital, and the owners of workers and factors of subsistence make demands and trade them in the form of wages and rents. They cede to the entrepreneur the ownership of goods that will have a higher value in the future, but that can only be acquired at the end of the production process. So both the supply and demand sides of capital goods depend on subjective perceptions of time preferences, which is the subjective theory of value promoted by Austrian scholars.

Digital Economic Value Theory

After the traditional economics theory of value and the Austrian economics theory of value have been discussed, we come to the digital economics theory of value. The focus will be placed on block chain technology and the block chain economy in order to discuss this issue. On the one hand, it is the transformative technological paradigm of the digital economy at this stage of its development, which has a reconfiguring effect on relations of production. On the other hand, block chain is the embodiment of a new generation of technological economy known as the "Internet of Value". The discussion of its value nature is equivalent to a discussion of the sources of value in the theory of digital economics.

Another name for block chain technology is the Internet of Value, and the reason for this is that it solves the following three basic problems of the original Internet. Firstly, block chain provides a solution to the flow of information about the flow of money (which is also known as the flow of capital) across the Internet through its application in the field of digital currencies. Secondly, blockchain solves the problem of authentication in the process of transactions through encryption and the reference to distributed ledgers. Thirdly, block chain identifies the problem of exchange of digital assets through the technology of consensus mechanism. But these three contributions actually answer only one question, which is how the element of value (price) in reality can be perfectly modeled and optimized by means of block chain technology. In short, how it can be bituminised by means of technology for a realistic economic or business model. These three contributions do not answer the question of the origin of the value of block chain technology. In summary, we need to ask the fundamental question of how its value is realized for the block chain economy?

In the field of the digital economy, block chain technology is now one of the most important technological paradigms. It is an important change and a major innovation for the development of finance not only in terms of the carrier technology for digital money, but also as an important underlying technology for the future development of the digital economy. However, there is a common sense that the evolution of a currency carrier is completely different from the carrying of credit. A currency secured by the credit of the state does not lose anything depending on whether the currency carrier is a shell or a banknote. At the same time, digital currencies carried on advanced block chain technology do not gain credit because of the complexity and innovation of the technology. These are the most basic common sense things we should know. So is the purpose of this article to discuss the worthlessness of block chain technology? No, we are trying to answer another question, its how can block chain achieve value? Simply put, how can block chain be made to live up to its title of "Internet of Value". More precisely, how do we understand the concept of value? In this chapter, this is the focus of the discussion.

The concept of value is discussed first. Classical economic theory is a theory in which value is the central topic, whereas modern economics is a theory in which price is the central topic. So the definition of value is to be understood from the work of the classical economists. My favourite classical economist is John Stuart Mill, who is also a genius thinker. He provides a very accurate and insightful description of the theory of value. In essence, Mill was a social philosopher. His open mind allowed him to explore and study classical economics through a number of disciplines, including philosophy, sociology and political science. At the same time, his research was the beginning of a new phase in the history of economic thought. Muller considered value to be "important felt". This explanation allows us to understand the meaning of value more precisely. We understand that in neoclassical economics, value is almost equivalent to price. But the fact is that everyone's preferences and behaviour are different, and the essence of this is that different things bring different "important felt" to each person. However, price is only a measure of this importance, and it is considered to be the embodiment of all value. This is clearly divorced from what we perceive as a real life scenario, and is a mathematical derivation of mechanical thinking that assumes all people to be rational.

The core of Mill's theory of value (also known as relative price theory) is his view that the purpose of value theory is to explain relative prices, and that absolute value is based on an invariant measure of value. In fact, what we usually judge as value is relative rather than absolute value, and supply and demand are established in specific scenarios. The 'important felt' value of the same good for a given person is different and leads to a different value. Therefore, we have chosen a framework based on behavioral economics to explain value, with the following three assumptions determining the value of a specific object. Firstly, the value of a specific object is inconsistent for a given individual, even for the same individual in different scenarios. So the magnitude of value cannot be judged in terms of price alone. Secondly, the magnitude of value is influenced by the judgement of the individual. More precisely, it is also influenced by the cognitive ability of the individual. Different people will bid differently on a block chain digital currency or stock due to differences in cognitive ability. Therefore, the price in this scenario becomes a reflection of the individual's cognitive ability. It is known from brain science research that there is a boundary between an individual's intelligence and cognitive ability, and that this boundary reflects behavioral consequences in terms of decision making or judgement. Finally, cognition is dependent on ability. Your cognitive ability leads to differences in the information you acquire, which leads you to make different judgements. This allows differences in individual decisions to emerge. In neoclassical economics, each individual's cognitive ability is assumed to be almost identical. Models based on Simon's behavioral economics are used to understand the differences in value judgements in real scenarios. Therefore, it is concluded that there are differences in each individual's abilities.

From the perspective of block chain technology, the basic elements of the three economies are gradually being built up by block chain technology. These changes are among the important elements that are endogenous to the business development process.

Firstly, the concept of community or community is re-identified with the basic scenario it offers to the transmission of value networks. An important branch of economics research is the study of group decision-making. That is to say, the externalities of business organizations are amplified through distributed network communities, which allow them to be repositioned in their dual identity as market and business. Most importantly, incentives that are different from the traditional development framework of the corporate system are provided through pass-throughs derived from block chain technology. We believe that this is one of the most important changes that the future block chain value network can bring to the development of the digital economy.

The second is the flow of information. In a block chain network, the flow of information is orderly and relatively transparent. So the phenomenon of change in decision making can be judged in the course of an efficient flow of information. Block chain technology is considered to be a "credit machine", which means that information in a block chain network is trustworthy and based on technical contracts. Therefore, the change in the flow of information refers to the change in value. On the one hand, the efficiency of the flow of value is increased by the high speed of information flow. On the other hand, transaction costs are reduced by the high speed of information flow. This is not only the infrastructure that block chain technology can be expected to provide for the digital economy in the future, but it is also the application scenario that is now the focus of the most mainstream innovators. In the future, an important foundation for the development of the digital economy will be the reduction of credit costs brought about by block chain.

Third, the establishment of a consensus mechanism provides an innate trust mechanism to the study of economics. The minimization of information asymmetry and the screening of consumers is a prerequisite for this mechanism. As a result, a deeper and more value-based business model can be built on it. In the context of such a business model, a business mechanism that is more participatory and can be customized to meet the needs of the user can be created.

The consensus mechanism is not only a technical mechanism, but it also offers the possibility that traditional business contracts are completely transformed into digital business contracts. And in the process, the contract is fulfilled by means of a smart contract. In short, the subjective behaviour of the market of the future means that contracts are traded and concluded, to a large extent, without the involvement of the traditional market, but through technological contracts. As a result, the efficiency of the entire market allocation of resources will be increased.

Such an ideal model and dynamic technological space is therefore highly attractive to researchers of the digital economy like me. If the boundaries of value are to be set for such a block chain economics, from which perspectives should one start? I therefore propose the following three basic dimensions to think about.

The first dimension needs to be thought of in terms of the value of information. Information is a fundamental element of block chain technology that has already been discussed. However, in the economic networks constituted by block chain, the right decisions require that sufficient information is available to the decision maker. At the same time, the decision maker has to make decisions on the basis of this information. In fact, the community of the blockchain can be understood as a society. The cognitive capacity of each individual determines the value of this society, which is also determined by the basic structure (institutional elements) of this society. The efficiency of the use of information depends on the boundaries of cognitive ability, while the efficiency of the dissemination of information depends on the structure or system. Therefore, if the value of a block chain community is to be judged, the dimension of information is the first criterion. Secondly, the pricing of information should depend on the value of information, not the cost of information. This is the basic starting point for the understanding of internet commerce or block chain business models. So one of the fundamental starting points is that the value of the block chain network is judged by the value of the information flowing in the network.

The second dimension needs to be thought of in terms of the value of the network. Block chain technology is part of a revolution and upgrade of the basic protocols of the Internet, and it is essential for the study of network economics. The study of block chain economics has to be analyzed from the perspective of the network, placing the individual within the group (community) and understanding the impact of the network structure on individual behaviour rather than isolating individual individuals. The study of networks is therefore important for the value of block chain technology. More precisely, quantitative heterogeneous utility in network economics should be studied. The study of the endogenous complexity of the network economy is fundamental to the study of network economics, so it is not the same as the pursuit of utility in traditional economics, which is the study of complexity utility (heterogeneous utility). It is simply explained by the fact that while in traditional economics utility is concerned with the outcome of choices, heterogeneous utility is concerned with processes. For each individual in the network, the concept of utility addresses impersonal needs, while heterogeneous utility addresses individual needs. So the ability of a block chain network to provide sufficiently satisfactory utility for each individual's individual needs is an important measure of the value of a block chain.

The third dimension needs to be considered from the perspective of mutual capital. Firstly, the concept of mutual capital needs to be understood as a specific concept in my theory of value in the digital economy. Based on the definition of the famous economist Fisher, anything that generates future income can be called a capital (which is also known as an asset). The definition of the Fisher equation means that "the price of an asset is equal to the sum of the discounted values of its future returns". However, the focus of our discussion is on mutual capital, which is all the elements present in a social network that facilitate cooperation in a prisoner's dilemma. The reason this concept covers the prisoner's dilemma in game theory is that the game in a network is very full and complex. So mutually beneficial capital needs to be used to measure these elements. In other words, in a block chain community, the value of capital is measured by the combination of the elements of "mutuality" that can be generated in the community after a full game. For the value of the network, these elements can be measured in terms of the discounted value of future returns. The logic of mutual capital is defined in terms of "mutual benefit" rather than "self-interest".

In digital economics, mutual benefit needs to be emphasised over self-interest. Adam. Smith famously wrote in The Wealth of Nations about the motives for economic behaviour, "Every man is constantly endeavoring to find the most advantageous use for the capital at his own disposal. He therefore considers his own interests rather than those of society. However, his study of his own interests will naturally or necessarily lead him to choose the most socially beneficial use." In other words, Smith believed that the interests of society as a whole were promoted by the pursuit of each individual's self-interest. The common synergy of the interests of society as a whole was realized through the unification of the individual interest with the common good in a commodity economy. This led to the traditional "first principle" of economics being that of self-interest.

I think that one of the very fundamental logic that has been consciously or unconsciously overlooked is that self-interested behaviour by individuals does not automatically lead to the realization of the interests of others or of society in fact. In other words, it is not self-interest that is a sufficient condition for the operation of a commodity economy. Since self-interest covers such a wide range of acts, the realization of many self-interested acts needs to be regulated by law. There are many self-interested acts in the commercial world that undermine business ethics and commercial rules, and the corresponding means of restraint need to be provided by the market and the law. In other words, only partly self-interested behaviour that meets the requirements of mutual benefit can promote the interests of society as a whole, and business is based on mutual benefit rather than self-interest. After people have become more efficient through the division of labour, they then share value through exchange. In a commercial society, the pursuit of self-interest by market behaviour has a boundary and a bottom line. This bottom line cannot go beyond mutual benefit. It is on this basis that normal market transactions exist. This logic of mutual benefit is considered to be at the heart of the value theory of digital economics, which means that value can only be created through mutual benefit and collaboration. The following three basic conclusions can be drawn.

First, value is created by cooperation and mutual benefit, not by labour or capital. In the development of human civilization, the most important reason for the triumph of homo sapiens, who were less physically conditioned and less able to work, over other species was that a common imagination of the future was built up through the richness of language. So a larger scale of mutually beneficial collaboration was created, which allowed the competition for the survival of the species to be won. After human beings have acquired mutual identity through imaginary communities, contracts and market mechanisms are established through mutual benefit. In economic life, it is the logic of the principle of mutual benefit that is most important, rather than the principle of self-interest that is traditionally emphasised in economics.

Secondly, human nature contains a selfish component is recognized and the value theory of digital economics is explored on this basis. All those factors present in social networks that facilitate cooperation in the prisoner's dilemma are mutually beneficial capital, premised on the fundamental problem studied in behavioral economics: the realization of cooperation. A mechanism for cooperation after full gaming is created by block chain networks. At the same time, the lowest cost consensus is achieved through a consensus mechanism. Only if the value network of the block chain and the value theory of digital economics are discussed on this basis can their content be understood. If economic growth through self-interest was emphasised in the industrial economy, the twin goals of income distribution and economic growth through mutual benefit are emphasised in the digital economy. The infrastructure provided by the digital economy is not only designed to solve the problem of efficient and low transaction cost growth of the economy, but also provides a solution based on a technological contract for the distribution of income following economic growth.

Thirdly, several fundamental balances are brought about by the value theory of mutual benefit. These balances are the basis for the understanding of the value theory of the digital economy. It covers three main balances, which are the balance between the interests of both parties to the transaction, the balance between costs and risks and the balance between the efficiency of both parties in completing the transaction. If the balance of interests between the parties does not exist, no transaction can be realized. The balance of values in a mutually beneficial state of personal subjective value dimension, recognized by both parties, is a prerequisite for commercial behaviour. If the balance of costs and risks does not exist, no sustainable transaction exists. Mutual benefit due to excessive risk is in fact a provocation to the underlying logic of business. If the balance of efficiency between the two parties does not exist, an imbalance of returns and an overall inefficiency of the market mechanism occurs, which allows the current market to be replaced by a more efficient one.

In summary, a basic framework of the elements of "value" in the block chain "Internet of Value" is provided. Three elements—information, network and mutual capital—are used to define its value. When the value of block chain is discussed, it is not only its technological innovations and concepts that should be looked at, but also its economic nature. If a block chain community or application scenario does not have the value of these three elements, then it does not actually have the attributes of a 'value internet'. If a measurement is confused by the "digital gold" packaged in block chain technology, and its heterogeneous nature in terms of economic properties is not understood, it is undoubtedly just an old trick of the "financial bubble". At the same time, the value theory of digital economics is based on the three basic value concepts of information, networks and mutual capital. Only when these three elements are understood simultaneously, the framework for thinking about value theory in digital economics can be established.

Chapter 7
Economic Thinking of Blockchain

In the previous chapter, the exploration of value theory in digital economics was presented through a discussion of the value of blockchain. Indeed, we argue that the intrinsic logic of value networks is possessed by the trusted network qualities in the technology of blockchain. The logic of this discussion is not only because blockchain technology has the specificity of the Internet of Value, but also because the most important change brought about by blockchain technology with the development of the information economy refers to the replacement of the commercial contractual regime by a technological contractual regime in order to allow the fundamental logic of the digital economy to be changed.

The mapping of the real world onto the bit world is realized by way of mathematics and algorithms. If this is the essence of the information economy that was once mentioned, then blockchain is the infrastructure that provides a richer mapping. So we can summarize the economic logic of the discussion of blockchain technology as a key outcome of the development of digital economics to date. In this talk, the main contributions of blockchain to the theory of digital economics will be discussed in the light of some fundamental features of blockchain technology, which are divided into three parts: open networks and the crypto-economy, consensus communities and complex networks, and platform and business ecology theory.

Open Network and Encryption Economy

Many researchers have paid attention to the emergence of blockchain technology, which appeared at two very important points in history, as follows.

The first historical point is the development of information technology, especially Internet technology: the history of the development of the Internet in the world shows that the APRA network was created in 1960 by the Advanced Research Projects Agency of the US Department of Defense, and it became the core of the development of the Internet. As of 2009, it was nearly 50 years old. With the help of

Z. Liu, *Principles of Digital Economics*, Contributions to Economics,
https://doi.org/10.1007/978-981-16-9020-4_7

German experts, China sent its first e-mail to the world on 20 September 1987, exactly 20 years after its launch in 2009. Over the last few decades, internet technology has evolved from its beginnings as an open network to a centralized structure with a few large companies at its core. From this perspective, blockchain technology is a re-opening of the internet.

The second historical point is the global impact of the world financial crisis in 2008, which triggered economic shocks in the global economic landscape. To date, the global economy is still affected by it. The first major application scenario for blockchain was the financial and monetary sector where credit problems were solved through decentralized technological mechanisms. We will also analyse the impact of the financial crisis that began in 2007 in depth in our subsequent discussion of global financial capitalism. The discontent of tech geeks with the global economic shocks and long-term income distribution caused by the financial sector and the proposed solutions based on blockchain technology need to be looked at here. This is not only a utopian ideal of a digital economy, but also encompasses the open source ideology and geek spirit that has been present since the late 1990s.

So blockchain technology is at a very important point of emergence for both the technology and financial industries. We don't think this is a coincidence. Blockchain technology is one of the most important fundamental technologies for the development of the digital economy. The analysis of the blockchain economy is therefore an integral part of digital economics. First, we analyse it from the perspective of the open network and the crypto economy.

In the beginning, the Internet network was created as an open platform to allow everyone to access information or benefit equally. However, over time the internet has become an increasingly 'monopolistic' network, which is not in line with the original purpose of its development. American cryptopunk thinking has influenced not only the development of the Internet, but also the development of blockchain. As Kevin Kelly in Out of Control [24], "the grassroots can suddenly regulate the otherwise arcane and forbidden realm of cryptography and code, leading to perhaps the most important result of all, the availability of electronic money". Blockchain technology's innovation in this area has been signalled for some time. Cyberpunk's technological ideas have been important to the development of the Internet and the development of blockchain technology.

So I think the roots of cryptoeconomics and blockchain technology did not emerge in 2009, but can be traced back to the Macy Conferences that continued to be held in New York in the 1850s, which covered a variety of topics and were initiated by Frank Fremont-Smith. Nearly 160 conferences were held in a cycle of nearly 20 years, involving people such as Wiener and Shannon. They were all scientists and thinkers who played a seminal role in the frontier science of their day. The main issues discussed included, but were not limited to, cybernetics, neuroscience, artificial intelligence and brain science, to name but a few. This series of conferences laid the foundation for human thought in the field of interdisciplinary research and the technological basis for the paradigm revolution in technology throughout the digital economy.

Next, we take a look at the changes brought about by blockchain, its cryptographic algorithms, distributed ledger technology based on consensus mechanisms, and creative solutions based on cryptopasses. When Bitcoin emerged, so did such cryptopasses. And it was accelerated across the board by the launch of Ethereum in 2014. In the open network design solution, the following three new and important technical features were made available to the traditional network economy by cryptopasses.

Firstly, decentralized network infrastructure is provided through distributed ledger technology to enable the openness of the network to be reconfigured so that the value of each node is revalued. Distributed ledger technology refers not only to new ways of circulating information provided by peer-to-peer communication technology based on encryption, but also to important technological innovations that transform the traditional double-entry ledger.

Secondly, the open network is incentivized through passes by different players, including investors, developers and users. They can be incentivized through the passwords to make the whole network work in a new economic model. Pass-throughs have the properties of ordinary money and the value of securitisation. In addition, the ability it has to circulate in the network based on usage rights is even more important, and it offers a new solution to the problem of the application scenario of the sharing economy in the digital economy.

Thirdly, the trend towards the centralization of the internet will be reversed through distributed ledgers and cryptographic passwords to ensure the openness, fairness and dynamism of the internet, thus allowing more innovation to emerge. It is worth noting that our view is not that a completely decentralized network is the best network, but that the rights of the participants in the network are better fed back to all parties involved in a model that is more in line with the spirit of consensus in the digital economy.

The above three technical qualities have led to applications like Bitcoin generating huge interest in the information technology sector and the traditional financial sector. On the one hand, the method of storing value is provided by Bitcoin through a decentralized network in a network. On the other hand, Bitcoin is a concrete application practice that offers a killer application of blockchain technology. However, Bitcoin should be noted for being only one application of blockchain and not the whole. Blockchain has a much greater value in that it enables the fundamental logic and ecology of the network economy to be reconfigured by way of cryptographic pass-throughs. Whereas in the past the network economy innovated through a venture capital-based approach, crypto-passwords drive innovation in the digital economy through a more open financing structure and innovative thinking. The launch of Ethernet for a network of smart contracts has led to the popularization of open networking technology for blockchain. While it brings with it great risk and regulation of innovation, crypto-passwords and open networks are also expected to develop in a more legal and compliant way in the future. Blockchain pass-throughs differ from the internet in that they embody the management of open network services and the provision of pathways to financing. It therefore has strong financial attributes. At the same time, the challenges and risks it encounters in the process of development are increasing.

Communities of Consensus and Complex Networks

After the crypto-economy has been discussed, the economics of decentralized community organizations (DAO) on the blockchain will be discussed in depth in this chapter. The main body of the blockchain's network economy is the community network itself, not the companies and markets, and its the conclusion we want to reach. Only when this issue is clearly understood can the nature of blockchain-based community organizations (DAO) be recognized and understood. It is also on this basis that the basic methodologies of community evaluation and construction can be understood and the value system of communities can be sorted out. The discussion is divided into three parts. Firstly, the economics of complexity is discussed. Secondly, the organization of complex networks is discussed. Finally, the economic logic of complex networks is discussed.

First, the theory of complex economics is discussed. This theory has already been mentioned before. At that time, the increasing returns quality of complex economies was mainly discussed. At this point we would like to present this theory in more detail. The theory was developed by Brian Arthur, a leading economist at the Santa Fe Institute in the USA. The theory was developed by Brian Arthur of the Santa Fe Institute. He is not only one of the most powerful economics professors at Stanford University, but also the winner of the Schumpeter Prize. He was not only a founding father of the Santa Fe Institute, but also a founding father of the science of complexity. Later, Arthur was awarded the first Lagrange Prize (awarded by the Royal Belgian Academy of Sciences for outstanding contributions to world mathematical knowledge) for his outstanding achievements. His main contribution was the development of the idea of complex economics based on the "law of increasing returns" through the study of positive feedback mechanisms. In other words, the economy is seen as a dynamic system that is constantly self-computing, self-created and self-renewing. Unlike neoclassical economics, which emphasises static resource allocation and general equilibrium theory, complex economics emphasises chance, uncertainty and the idea that "all change is possible". It is a dynamic economic discipline based on prediction, reaction, innovation and substitution.

The nature of complexity economics leads me to believe that any understanding of the digital economy should be based on the paradigm of complexity economics, and in particular on the economy of networks (which includes the internet and blockchain), otherwise its nature cannot be understood. The theoretical background will not be covered in detail due to space constraints (I recommend Brian Arthur's Complexity and the Economy [25]). In complexity economics, the economy is considered to be an unbalanced system, and this theory will be discussed. However, as the economic world becomes more and more uncertain, and as complexity economics is based on the idea of biological evolution rather than a Newtonian continuum, a more open and dynamic system of inclusion is used to view changes in the economy. As a result, complexity economics is slowly moving closer to the centre of economics and, in particular, of the research theory of the digital economy. It is only when the existence of disequilibrium (which is also known as dynamic equilibrium) is recognized through the paradigm of complexity economics that issues of innovation, evolution and value in digital economics can be addressed and understood. Thus, digital economics will be defined by me as the "economics of the innovator".

Second, complex networks are discussed. In blockchain, consensus community organization can be defined as a complex network according to complexity economics. The concept of community network organization must therefore be understood from the perspective of network economics and complexity economics, which refers in particular to the difference between networks and markets and firms. In fact, there are two basic misconceptions about networks. Firstly, the concepts of network and market have been conflated so that the logic of the market economy is used to discuss the network economy. Secondly, the notion of network and enterprise has been conflated in such a way that the differences between network communities and enterprises cannot be understood. These two misconceptions have not only caused many blockchain innovators to lose control and fail in the management and operation of community organizations, but also prevented investors in projects in the blockchain space from properly grasping the value of their projects. So the relevant conclusions need to be sorted out theoretically.

Finally, the economic logic of complex networks is discussed, and it is divided into three main areas as follows.

Firstly, the differences between the network economy and the traditional industrial economy are discussed. If the economic environment of the industrial era is compared with that of the network era, the biggest difference is the homogenization of the industrial economy. Simply put, in an economic system the factor demands are essentially the same and the consumer needs are essentially the same. So the economic system forms a network of rules. The network economy, on the other hand, is heterogeneous and refers to the diversity of demands of the economic system, which is full of uncertainty. So what is formed in the network economy is a chaotic economic system. The following two basic models of the network economy are presented here. One is the small world network model defined by Duncan Watt and the applied mathematician Stogatz in 1998. The other is the scale-free network model proposed by Albert Laszlo Barabasi, a leading researcher in network theory who wrote "Links". The scale-free network model was proposed by Albert Barabasi, who is not only the director of the Centre for Complex Networks Research at Northeastern University, but also the founder of the Network Science Society.

Secondly, the increased efficiency of economic operations and the significant reduction in transaction costs under the logic of these two network economic models are discussed. Information transparency and trust relationships between acquaintances are used to model the new economy with its logic of maximum optimization for small world networks. Transaction costs are reduced while equal efficiency is achieved. This is primarily an economic model of stochasticity. In contrast, a free economic order based on complex networks is constructed by scale-free networks. On the one hand, this order has the characteristics of a free economic order with random networks. On the other hand, it possesses the efficiency advantages of a rule-based network economy. When all organizational groups (households, markets, firms and networks in the narrow sense) are considered to be part of a wider network, the scale-free network model is the most complex and well-established economic model in the organization of these network models. In other words, a basic logic is explained by the small-world network model, which refers to the reduction of market transaction costs due to the current network economic efficiency in the digital economy. In short, transaction costs are reduced through trust and transparency of information. The basic logic of the future blockchain network economy is explained by the scale-free network, which refers to the construction of a free economic order based on generalizations (which corresponds to the distributed network of the blockchain economy model) and a distributed organizational order based on technical contracts (which corresponds to the consensus mechanisms and smart contracts of the blockchain economy).

Finally, we discuss consensus-based decentralized community organization structures. Blockchain consensus-based community organizations (DAO) will be the focus of discussion. The blockchain community network structure is understood to be based on the rules between the free economic order and the contractual economic order after which the community network can be understood as an integrated structure that has not only the efficiency of free transactions of a market but also the function of a business to fulfil specific contracts. A trust relationship based on consensus is reached by the blockchain network based on peer-to-peer distributed contracts. On the one hand, the stranger relationship of the market economy is transcended and transaction costs are reduced. On the other hand, the boundaries of the community networks based on acquaintanceship in the original Internet are expanded. These two aspects reduce the transaction costs of DAO in the blockchain economy to a level below that of the traditional market economy, and even below the basic academic logic of the existing internet economy. In the context of consensus mechanisms and distributed networks, DAOs become a kind of intermediate existence between industry and business. On the one hand, efficient distribution of capital is achieved in the form of distributed networks, and information asymmetries are resolved. On the other hand, a credit system that goes beyond simple acquaintanceship is established through consensus mechanisms to enable a new economic model in the form of the sharing economy (also known as the sharing economy) to emerge, allowing a series of restrictions on the relationship between firms and markets in the neoclassical economy to be broken down and new organizational structures to be created.

In summary, the phenomenon of complex networks in complexity economics is discussed. Decentralized consensus communities in blockchain are defined as complex network organizations. So two basic economic models of networks are introduced, which include the small-world network model and the scale-free network model. Finally, the extremely high efficiency and low transaction costs of this economic model of complex network organization are discussed, which is the basis for the subsequent discussion of smart contract theory and the institutional economics theory of blockchain.

Platform and Business Ecology Theory

With the complex network organization being discussed, we can't help but wonder how such an organizational structure would be defined? I am not a fan of economists who invent new concepts. I think that concepts and definitions of most things have already emerged in the course of business research. But the fact is that we have already been exposed to the relevant theories in the existing digital economy. The structure of a network is a combination of a random spontaneous order and a contract-based order, so the organizational structure of a community network should also be a combination of the two. And platforms and business ecologies are the two basic models that fit this combination.

In the Internet economy, platforms denote network organizations formed by various bilateral economic models, such as e-commerce platforms, content platforms and social platforms, among others (e.g. Jingdong, Toutiao, Weibo, etc.). A large-scale network for meeting customized needs is provided by such platform-based organizations, which form an economic model with a distinct centre as a prominent feature. The cost of the platform is reduced by this platform organization through a mass customization of services and functions similar to that of the industrial era, and its essential other functions are differentiated through an infrastructure that is provided on the platform to form differentiated services and customized products. In the platforms of the Internet age, exponential growth is not only achieved, but the advantage of monopolistic competition is gained. At the same time, it has brought about huge changes to the entire business ecology of the digital economy, which is reflected in the following three main aspects.

First, consumer-focused models are being created by platforms. The sharing economy and C2B are both business models that focus on the enhancement of user experience and service. Data on the needs and behaviors of consumers is made available by big data in networks, and its led to platforms becoming a business model in which consumers define products and services in order to enable users' value propositions to be quickly aligned with the platform's providers.

Secondly, the gradual convergence of the boundaries between consumption and production is being realized by platforms through digitalization. It has not only led to lower transaction costs, but a digital economy with services and experiences at its core has more flexibility in terms of factor combinations. At the same time, entrepreneurs in the Internet age are equipped with an industrial environment that liberates productivity as well as allowing them more freedom to choose their space.

Finally, it is observed that a network-centred marketplace is constructed by the platform. In this marketplace, all infrastructure is provided by the platform, and all participants and elements are traded in the platform so that the expansion of market boundaries and the construction of a networked organizational ecology are realized.

The business ecosystem is built as a decentralized organizational model, and the business ecosystem forms an economic model based on the emergence and generation of incremental rewards. Simply put, the economic elements in a business ecology become shared instead of exclusive, and different contractual mechanisms are formed by consensus among the parties in the ecology. The associated processes of trading, bookkeeping, dissemination and value distribution can be automated by these contractual mechanisms through smart contracts. Due to space constraints, the theoretical model of business ecology will not be discussed in detail here. However, it is straightforward to conclude that the competition of community organizations in the future represents the competition of business ecologies and value networks, where different business ecologies are based on different consensus (value propositions) and different value networks (blockchain community DAO) are composed of different value propositions. Therefore, the future competition refers to the competition between dynamically linked business ecosystems. For example, the current competitive dynamics of BAT is essentially a business ecosystem rather than a single market competition. At the same time, undifferentiated competition and the reshaping of value networks are being unleashed in a number of areas, which are the qualities of business ecology competition.

In the Internet era, the concept of business ecology is valued. The most important economic element of the online business ecology is the data generated by the online behaviour of linked users of the Internet. The differences between the connotations of business ecology in the traditional industry, the Internet and the blockchain can be interpreted from the following perspectives.

Firstly, the traditional concept of business ecology refers to the holistic and systematic logic of business being understood through the concept of ecosystems in ecology. This concept was first introduced by the academic Moore in 1993 and he explores it in depth in "The Decline of Competition" [26] in 1996. He argues that the system of firms, customers, suppliers, major producers, competitors and other stakeholders in traditional industries is analyzed primarily by business ecology. And this analysis of the equilibrium evolutionary process of the system can be useful for the formulation of a firm's strategy. Firms have different ecosystem-level challenges at different stages, so firms need to adopt different strategies at different stages. Businesses within a business ecosystem should aim to create new wealth so that new ecological cycles are created, which can replace a narrowly based industry competition strategy.

The second is the concept of business ecology in the Internet era. 1998 saw scholars Gossain and Kandiah cite Moore's concept of business ecosystems to discuss the Internet, leading to the introduction of an organizational model that links all market participants. They argued that in the network economy constructed by the Internet, organizations can link deeply and exchange resources with each other through the network. In 2004, the business ecosystem was studied in depth by Lansiti and Levien in terms of the biological ecosystem theory. They defined a business ecosystem as "a group of interconnected, value-creating and value-sharing enterprises". In the Internet age, this is the basic concept of business ecology as understood by most researchers.

Thirdly, the concept of business ecosystem constructed by blockchain network is different from the other two for the reason of the technical characteristics of blockchain itself and its qualities as the Internet of value. On the one hand, the business ecosystem of blockchain encompasses the forms and connotations of networking as well as self-organization in the business ecosystem of the Internet. Transaction costs are reduced through the networked structure and the proliferation of business boundaries is achieved through a self-organizing ecology. On the other hand, the barriers to competition within the ecosystem are locked in by the blockchain network through the cryptographic economy, and are redefined through property relations, which allow for a new system of governance of the business ecosystem, thus extending the value of productive relations beyond transactional relations. This is the key difference between the blockchain ecology and the traditional Internet ecology.

In fact, business is understood based on the concept of ecology. The term competition is not really accurate in this context. In the future, there will not be a concept of "competition" as it is understood in the commercial marketplace (which means that market space is gained by benchmarking rivals and defeating competing products). Every business ecosystem is constantly evolving. In the process of evolution, it becomes more viable and adaptable, with more efficient value networks and more valuable products and services, in order to survive (which means winning the competition). In other words, the idea of evolution should be used to look at the relationship between the business ecologies of the future. In a business ecology, the winners who provide services and products survive, while the losers naturally disappear. This is what is happening in the business ecology. It means that you are eliminated not because of the strength of your competitors, but simply because of your own unsuitability. And the harsher thing is that the more you excel in the business ecology, the faster you are eliminated. This is the nature of competition based on an innovative networked economy. This natural selection process in a brutal competitive environment is akin to the kind of war between civilizations mentioned by Liu Cixin in "The Three-Body Problem". A silent, survival-oriented and uncertain evolutionary process is taking place between business ecologies. It is pointless to focus on competitors. This is the truth about organizational competition in the digital economy, because the entire business ecology can be eliminated.

Finally, here are the points that have been highlighted by us. Digital economics is a growing and iterative discipline. At the same time, the economic phenomena we observe are constantly evolving. Therefore an evolutionary mindset should be used to view economic phenomena rather than in a static way.

Chapter 8
Blockchain and Austrian Economics

In this chapter, the basic historical origins, research methods and influence of the Austrian School of Economics on digital economics will be discussed in detail, especially as the inner logic of this influence is explored through the study of the blockchain economic paradigm.

Austrian School of Economics has a very special historical position and theoretical connotations compared to mainstream economics.

From the point of view of theoretical identity, the subjective theory of value and the market process theory are the foundations of Austrian School of Economics, and a theoretical system of objective economics is constructed that differs from the neoclassical one. From the point of view of economic thought, a set of logics of economic analysis that differed from the "rational economic man" assumption of neoclassical economics was constructed by Austrian School of Economics on the basis of Mises' idea of "human action". From the point of view of ideological tendencies, government intervention in macroeconomic policy is strongly opposed by Austrian economists. At the same time, they argue that the cause of inflation and the fall in interest rates is government intervention in monetary policy, which distorts price signals and generates economic cycles.

In fact, a group of economists who have had a major impact on the development of economics belong to the Austrian school, including Menger, Bohm-Bawerk, Wiesel, Mises, and, as we have mentioned, Joseph Alois Schumpeter and Friedrich Hayekk. In addition, the roots of a number of important topics that have been and will be discussed in digital economics can be found in the work of Mises and Hayek, for example, information asymmetries, the new institutional economics, the evolutionary theory of economics and the idea of property rights.

This is why it is so important to understand the historical origins of the Austrian school of economics, its methodology and the links between it and digital economics. On the one hand, this can lead us to the intellectual roots and basic methods of the theory of digital economics. On the other hand, when we discuss other topics in the future, this can help us to establish the basic research framework and methods to

Z. Liu, *Principles of Digital Economics*, Contributions to Economics,
https://doi.org/10.1007/978-981-16-9020-4_8

understand the methodology of the discipline's cognitive world and the particularities of economists.

Due to space constraints, the following three basic topics will be briefly discussed, including the origins of the Austrian School of Economics, the methodology of the Austrian School of Economics, and algorithmic economics and Austrian economics.

Origin of the Austrian School of Economics

In the early period, three scholars belonged to the Austrian school of economic thought: Menger, Wiesel and Bohm-Bawerk. The main figures of this school of thought were Mises, Hayek and Rahman. The contributions of Menger and Mises will be highlighted here, while the views of the other scholars will be dealt with in the following chapters.

Firstly, the Austrian economist Carl Menger will be discussed. We mentioned him before. He published his "Principles of Economics" [27] in 1871. In this book, the foundations of classical economics were completely overturned with its reference to the objective theory of value. And a systematic subjective theory of value for economics was pioneered. The present principle of marginal utility was discovered by Menger at almost the same time as William Stanley Jevons in England and Leon Walras in France. So they also belong to the founders of modern marginal utility. Therefore, Menger was the first to emerge as the founder of marginal utility in neoclassical economics. But his framework of economic analysis differs from that of the other two economists in that he was the first economist to systematically present a framework covering "individualism, subjectivism and marginalism". In fact, mainstream economics (neoclassical economics) has followed two paths to which we have referred. The first path refers to the use of mathematical forms of abstract reasoning, which is the path of Walras. The second path refers to the emphasis on the use of statistical processes to test theoretical hypotheses, the path suggested by Jevons. In addition, Menger pioneered a third path, which refers to an economics based on the subjective study of economic processes and institutional theory at its core. This is also known as the Austrian school of economics.

If subjective economics is considered to be economics constructed on the basis of formalistic mathematical ideas, the Austrian school of economics is considered to have developed in a non-mathematical way. Law and institutions were integrated into its analysis, and economics-related studies of institutions, processes and disequilibria were focused on. Menger's approach to economics is a subjectivist approach to economics, which focuses on 'organic' institutions. Simply put, while it is a product of human behaviour, it is not the institutions that humans are involved in, such as money, markets, language, communities and law. The most important issue in economics is defined by him as the origin and evolution of "organic" institutions, rather than "pragmatic institutions".

If it is traced back to its roots, all these elements have been included, according to the present study of the history of economic thought, almost exclusively in the tradition of the discourses of the Spanish scholastic on economic questions. Menger was merely the first to systematically present an entire system of theoretical economics containing these elements. Wiesel and Bohm-Bawerk were disciples of Menger. The first economist to use the term "marginal" was Wiesel. The Austrian school's theory of interest and capital was expanded by Bohm-Bawerk. This was the earliest origin of Austrian economics.

Some of Menger's fundamental reflections in the course of his research must therefore not go unmentioned. These reflections not only belong to the qualities of the Austrian school of economics, but are also at the root of important matters in the understanding of digital economics. Three of these reflections will be briefly introduced.

First, Menger was the originator of the marginal principle, a viewpoint concept that was not restricted to continuous and differentiable marginal utilities was developed by him. He considered that mathematical methods were completely inadequate for the understanding of human behaviour to obtain the essence of the economic phenomena whose explanation he was attempting. This led the Austrian school of economics to avoid as much as possible the methodology of using mathematical tools to explain economic phenomena. As Menger puts it, "How can we gain knowledge of essences, such as the nature of value, the nature of land rents, the nature of entrepreneurial profit, the nature of the division of labour and the nature of the compound standard, etc., by means of mathematical methods?" Menger's attitude towards the expression of mathematical formalism was one of total opposition, and its led to the fact that Austrian economics did not take mathematics seriously.

Secondly, a series of frontier issues that are still being studied by economists were raised by Menger. He had an insight that no other economist of his time had, covering uncertainty, the behavioral biases of people, the non-determinism of phenomena and the asymmetry of information. The focus on these issues, in turn, turned out to be a process of concern for the Austrian school of economics, rather than the ideological roots of equilibrium analysis theory. The theorists of neoclassical economics, with their obsession with computable economic theory, deduced an entire theoretical edifice of economic research through the simple rational man hypothesis, which resulted in a neglect of the individual and a lack of explanatory power for macroeconomic theory.

Thirdly, Menger's idea of subjectivism is not only a dynamic subjectivism, but it is also radical. People's economic behaviour is not simply restricted to personal preferences. Instead, it is extended to concepts such as people's expectations, costs as well as time. As a result, economic actors are considered to be active and adaptive objects of their environment, rather than passive reflectors as in neoclassical economics. In fact, this is more in line with the laws of human behaviour in our reality. Interest, for example, is considered by Austrian scholars to stem from human time preferences. Any production takes time. Therefore, interest rates need to be taken into account when producers, owners of resources and consumers make decisions.

Such an idea is derived from the subjective theory of value and the consideration of "human action".

Methodology of the Austrian School of Economics

The original origins of the Austrian school of economics were discussed above. In fact, the 'modern Austrian school' denotes the Austrian school of economics based on human action, pioneered by Ludwig von Mises, who was a student of Bohm-Bawerk. Murray Rothbard and Friedrich Hayek. Hayek are two of his most famous students. Here, the contributions of Mises and Hayek are highlighted, especially the basic methodology on the study of modern Austrian economics that Mises constructed in his economics work Human Action [28]. Of course, Hayek's methodology on the study of the Austrian school of economics will also be discussed and it will be used as a reference and a comparative way of thinking.

Firstly, the basic ideas of human behaviour will be discussed. Mises believes that the basis of all economics is that individuals engage in actions that have a specific purpose. Simply put, people achieve their goals through specific actions and its the basic logic for understanding economics. Everyone has basic goals, so everyone will have different values. At the same time, different values have different means of adoption, so each person acquires different technical knowledge. And whether a person can achieve his or her goals depends on different knowledge and means. Therefore, the correctness of one's valued goals will not be judged by human behavioral science, which judges whether each person has clear values followed by specific means to accomplish the goals. This is the root of all economics. All economic behaviour derived by Mises on this basis takes time, and all behaviour is characterized by uncertainty. Moreover, the means to reach the goal is scarce (if it were not scarce, the goal would be reached immediately. Then the action loses its meaning).

So an economic model based on human behaviour is derived by Mises through a very clear and rigorous logic. In this economic model, the values, goals and content of human actions are not studied. Instead, only people's aims and the behaviors they act upon in order to achieve their goals are attended to. The explanation of the concepts of utility, demand, supply and price, etc. in economics is then generated by these behavioral traits. Menger's ideas, which we have just mentioned in terms of the analysis of the subjective values and choices of individuals, and the opposition to econometrics are also extended by this methodology. Whereas the natural sciences are based on the empirical discovery of invariant quantitative relationships, econometrics cannot identify an invariant relationship in human history, as Mises repeatedly emphasises. Human behaviour scholars, on the other hand, believe that individuals think and that they act on their own values and expectations. This allows the homogenized utility of neoclassical economics to be avoided, and the behaviour of each individual to be given real attention.

Next, Hayek's contribution will be discussed. Although both Hayek and Mises were important scholars of Austrian economics, their theories still differed significantly, particularly in their understanding of subjectivism. Abstract rationality was the rationality that Mises emphasised, while institutional rationality was what Hayek emphasised. Two of the greatest economists throughout the twentieth century were Hayek and Keynes. The approach to the world economic landscape today is also being influenced by their theories. The Great Depression of 1929 and the global financial crisis of 2008 can be explained by Hayek's business cycle theory. Keynesian behaviour against national governments refers to economic development being driven through the constant quantification of loose monetary policy and deficit fiscal policy. Hayek's theory has had a very broad influence. Among the Austrian economists, Hayek was the most thorough researcher of Austrian liberal economic thought and public policy. He is also by far the most influential of the Austrian economists.

Finally, the classic methodology of the Austrian school of economics will be discussed, with its reference to the process theory of markets. The market is emphasised by it as a process, rather than a set of mutually coordinated elements consisting of price, quality and quantity, which produce a state of market equilibrium. This is quite different from the idea of perfect competition and equilibrium as emphasised in neoclassical economics. At the same time, it is the basic starting point for the theory of dynamic equilibrium based on network organization in the subsequent discussion. In the Austrian school of economics, the role of knowledge and expectations is emphasised in relation to the achievement of equilibrium. A dynamic theory of the market process needs to be constructed and the systematic changes in information and expectations by virtue of which all market participants act need to be explicitly considered as a guide to the evolution of the market structure.

In 1988, Hayek published his last book, "The Fatal Conceit". In this book, the ideas of contemporary evolutionary social theory are referred to as being derived from the implications of market process theory. And the basic ideas of evolutionary social theory are said to be constructed from research on the evolutionary process of behavioral rule systems. In this context, the structural evolutionary theory of complex economics that we discussed should come to mind. This reflects the influence of the ideas of the Austrian school of economics. Schumpeter and Bryan. Arthur were both clearly influenced by this idea.

For process-based economic theory, it is important to note that the essence of market-based process theory is human-based economics. In the concept of human-based action constructed by Mises, people are more than simply people who rely on rational calculation. People who are alert to chance and contingency act when they are presented with opportunities that enable them to achieve their ends. In classical economics, the human being is understood simply as the actor who allocates resources after encountering an opportunity. The motivation and process through which this person recognises the opportunity is left unexplained. Thus, this vigilant person who constantly finds opportunities actually has the ingredients to become an entrepreneur, whose process of referring to resource allocation decisions is transformed into a cognition about human action. The theory of market equilibrium is transformed by the entrepreneur into a theory of the market process. Schumpeter, Knight and Mises, among other scholars of the Austrian school of economics, had the important role of emphasising the role of the entrepreneur.

Algorithmic Economics and Austrian Economics

Once the value and thinking of Austrian economics to the digital economy field is understood, the relationship between algorithmic economics and Austrian theory will be discussed. In other words, the problems solved by Austrian economics within the framework of algorithmic economic theory are discussed. Although the concept of algorithmic economics is not widespread, almost all research in the field of computing has been influenced by the concept and ideas of "algorithmic economics". The idea of the algorithmic economy has influenced both the crypto-economy, which we have mentioned, and the sharing economy, which we will discuss in the future. The algorithmic economy will be analyzed as an important module in the framework of digital economics. This will not only allow the impact of algorithmic economy ideas to be understood, but also the underlying theory of digital economics in terms of the convergence of technology and economics.

First, the concept and logic of algorithmic economy will be discussed. This has its origins in the classic definition of a program by the Turing General winner Pascal, which refers to a program = algorithm + data structure. It is the most classical definition of a program in the field of computing. This formula has sufficient explanatory power for technological concepts such as the internet, big data, artificial intelligence and blockchain (distributed ledger). All information technologies are built according to the logic of "algorithm + data". Algorithms can be considered as the most important connotation of "numbers" in the digital economy. Algorithmic economy means an economic model in which the logic and rules of economics and business are "algorithmicised" and then put into computer code to be implemented automatically. In the traditional economic sphere, the market and economic logic, which is achieved by matching supply and demand with the "invisible hand", is greatly enhanced by algorithms and technology, resulting in a significant increase in economic efficiency and a reduction in transaction costs. This is what the algorithmic economy is all about. To date, the most mainstream applications of the algorithmic economy are the crypto-economy and the sharing economy. The former promotes distributed collaborative production between enterprises through incentive-compatible algorithmic rules and contractual arrangements in the ecology. The latter achieves cost reduction and large-scale market scale expansion through usage-based value sharing.

In fact, a different logic to that of resource allocation in traditional economics is provided by algorithmic economic theory. The market economy enables the circulation and allocation of resources through the price mechanism. A solution that combines the attributes of planning with those of the free market is provided by algorithmic economic theory. In economist Ronald Coase's "The Nature of the

Firm", the theory of transaction costs is discussed. He argues that the transaction costs of organizing the division of labour in the firm are lower than the costs of organizing the division of labour in the market leading to the creation of the firm. So the enterprise belongs to the market as a kind of organization of resource allocation, the underlying logic of which is the high or low transaction costs. In addition, in the digital economy, new organizational forms are offered, which refer to the sharing economy platforms represented by networked organizations and the Decentralized Autonomous Organization based on blockchain technology (it is abbreviated as DAO). In the framework of the algorithmic economy, it can be analyzed from the point of view of transaction costs; the institutional costs of a Decentralized Autonomous Organization are lower than those of a traditional enterprise, which means that a more compatible and scalable resource allocation mechanism can be provided through the algorithmic economy, in addition to the enterprise and the market.

Austrian economics is explicitly opposed to planned economies. In both Mises' Socialism: An Economic and Sociological Analysis and Hayek's The Road to Slavery, government intervention is strongly opposed. In the field of the digital economy, a political theoretical framework in which market economy and interventionism are compatible is proposed by algorithmic economic theory, whose basic logic is similar to that of the Polish economist Oskar Lange, who lived in the US. The underlying logic is similar to that of Polish economist Oskar Lange's theory of market socialism. This theoretical framework and the realization of the algorithmic economy as a reconciliation between liberal order and interventionism will be discussed here.

First, the position of Austrian economists on political economy is discussed. The economist Huerto de Soto is the most representative of the modern Austrian school of economics. According to him, only two alternative systems exist in theory and time. It means that people either fully enjoy the freedom to exercise their entrepreneurial talents (in a context where there are no restrictions other than the minimum number of traditional criminal and civil law rules required to recognize and protect the private ownership of the means of production and to protect human behaviour and contracts from systematic infringement) or that in almost all markets and broad areas of society, entrepreneurial talents are subject to systematic and general coercion, and in particular private ownership of the means of production would be impeded. There is no doubt that this view is the dominant one among Austrian economists. Based on the derivation of Mises' theory, both social coordination and economic calculation can only take place within a system in which human behaviour can be fully exercised. They argue that the fundamental error of Keynesian economics is a departure from the real economic world.

Austrian economists believe that in the real economic world there is no aggregate demand, aggregate supply or overall level of output or employment in an economy. They are constructed by economists or statisticians to describe economic phenomena. In the real world, only different differentiated demands and differentiated products and services provided by different business organizations exist. The market mechanism is based on the free action of such different people, which is the basic logic of the entire socio-economic system. According to this theoretical logic, capitalism and the free market are the driving force of human prosperity, while government intervention is the stumbling block. This is the theory derived by the Austrians. Austrian economists argue that the over-reliance on government power in public policy being used to secure and maintain economic prosperity and stability is difficult to be achieved in the long term. Next, we will discuss the question of whether there is really no economic theory that combines socialism and the free market by means of regulation and intervention. Such a logic can be realized, at least in the field of algorithmic economics.

Next, the idea of market socialism will be discussed, including the basic logic of this theory and a critique of Mises' theory. Due to space constraints, only Oscar Lange's concept of market socialism is discussed. The concept of market socialism as proposed by Oskar Lange is discussed. This is because it breaks with the traditional socialist understanding of planning and the market for the first time and introduces the market mechanism into the socialist economic system.

The economist Baroni's idea of general equilibrium was referenced by Lange. This idea is mainly an extension of the Pareto of the Lausanne school, which is concentrated in Baroni's "The Ministry of Production in the Collectivist State", published in 1908. The main aim of this essay was to use mathematical methods to argue for the way in which a socialist production administration should direct production in order to achieve the optimum allocation of resources. Baroni argues that prices in the capitalist system are simply an exchange function. Socialism can use this exchange function to maximize the welfare of consumers and the utilization of scarce resources in order to achieve general equilibrium, while ensuring that the marginal rate of utility of consumer goods is the same for all consumers and the marginal utilization of scarce resources is the same for all producers. Lange argues that a socialist economy can be economically accounted for through broad-based prices. In a socialist society, an order of priorities guiding the choice of action can be formed and the amount of available resources can be mastered. This is achieved through the technical possibility of changing one commodity into another, which refers to the production function. Of course, other conditions of choice can also be known and provided. On the basis of these three sources, socialism is perfectly capable of achieving rational economic calculation and ultimately general equilibrium within the context of the commons.

Lange makes direct use of this result of Baroni in presenting the basic logic of market socialism. At the same time, Lange also drew on Taylor's trial-and-error approach. He argued that as long as the socialist state could guarantee a certain monetary income to the citizen and authorize him to spend this income, he could selectively buy the goods produced by the state. Thus, socialist economic equilibrium could be achieved through the constant observation of market conditions by those in power and the timely adjustment of the prices of various factors. This clearly illustrates the specific way in which socialism achieves equilibrium, i.e. the trial-and-error method. Lange prefaces his vision with a brief criticism of Mises's view. In his view, the reason why Mises asserted that a socialist economy could not solve the problem of rational allocation of resources was that he did not correctly understand the nature of prices. Vickers distinguished between prices in the narrow sense and prices in the broad sense, and Lange argued that this distinction between the connotations of the concept of price would refute Mises' assertion.

Finally, programmable society and algorithmic economy ideas are discussed. Once Lange's ideas are understood, their feasibility in other economic models will not be discussed, while the feasibility of such a model in the digital economy is discussed. The basic model of neoclassical economics that has been discussed is criticized on the grounds that real-world economic factors cannot be realized simultaneously as embodied in the equations. Real-world market processes are very cumbersome and complex. However, it is true that all transactions in the digital economy can occur simultaneously, and the technological basis for this is provided by blockchain and algorithmic economics.

Let's start our discussion with the distributed ledger, a mechanism for centralized computing (consensus mechanism) is provided by the distributed ledger. This mechanism is responsible for managing and formulating the logic of all transactions in the blockchain network, and the construction of an intervening market mechanism is achieved through the organization, recording and verification of the identities, powers and obligations of all participants. However, once the relevant transaction mechanisms and computational economic models have been developed, all nodes operate according to the order of the free economic market. This model then fulfills a double management function, which means that the intervention in the overall economic market and the setting of the basic economic logic is achieved through management information. This is the reason why the design of the pass-through economic system is discussed in the context of blockchain networks. On the other hand, the availability of a market network with full transparency of information and no possibility of rent-seeking behaviour is used to incentivise innovation by all parties involved in the network to ensure the positive evolution of the economic ecosystem.

Finally, an understanding of algorithmic economic thinking based on contract theory is discussed. The process of contract making and performance is the essence of the market. Transactions are realized through the making of contracts. In the course of a transaction, transaction costs refer to the costs incurred by the participants in realizing the contract (including search costs, negotiation costs, contracting costs, etc.). The existence of transaction costs is predicated on the uncertainty of markets and the imperfection of information. From Coase's law, it follows that firms are founded on the premise that lower transaction costs are achieved by replacing short-term contracts with long-term ones. However, the long term contracts made by the firm are still imperfect as there are a series of ex-post transaction costs and enforcement costs of the contract. A range of technical contractual solutions (e.g. smart contracts) are provided by the idea of algorithmic economics, which aims to achieve reciprocity of rights and obligations in all states of the transaction process, and to record all information during the transaction. It possesses the perfection of Turing. In short, a better solution than the long-term contract of the firm is offered by the algorithmic economy, which allows the dilemma of rising transaction costs due to the increase in the size of the firm to be solved, so that the economic model of optimal resource allocation is realized algorithmically.

In summary, the relationship between blockchain and Austrian economics was discussed in three parts: the origins of the Austrian school of economics, the methodology of the Austrian school of economics, and algorithmic economics and Austrian economics. In earlier, we introduced you to the evolution of Austrian economics and its basic methodology. In the latter parts, we analyzed Austrian economics' view on the mechanisms of an intervening market economy and the logic of market socialism as proposed by Lange, and propose a framework for solving the problem of information transparency and computability through the mechanisms of distributed ledgers through the analysis of transaction costs and contractual regimes in the algorithmic economy. This is a way of thinking about the relationship between technology, economics and social institutions through an interdisciplinary approach to our understanding.

Chapter 9
Network Organization and Innovation Management

Previously, entrepreneurial theory and innovation was discussed, with its focus on the importance of entrepreneurs from an economic perspective and how entrepreneurs can drive technological iteration through innovation, thus enabling dynamic and balanced economic development. In this chapter, the topic of innovation management in networked organizations in the digital economy will be discussed, which also means that innovation is understood from the perspective of business theory and management theory. Here are the reasons why this topic is discussed.

Firstly, the main form of future business organization is the networked organization in the era of the networked economy (here, the concept of enterprise is used for the time being to illustrate this). It is therefore very important that the inner logic of networked organizations is clarified, especially with regard to the changes in the inner evolutionary logic of the organizational ecology of the enterprise in the digital economy.

Secondly, the core driver of economic development is to drive the application of new technologies through organizational innovation in order to reshape the value network in the digital economy. Therefore, the basic theories and mechanisms of innovation management in networked organizations need to be understood. Here, the model and ecology of open innovation needs to be discussed. In network organizations, open innovation is one of the most important models of innovation. In other words, innovation in SME is realized through the coupling of inward innovation from the outside and outward innovation from the inside.

Finally, business organizations exist in the role of builders or participants of platforms or ecologies in the future of the digital economy. Thus, an understanding of the management theory of networked organizations is based on how the strategies of networked ecologies are constructed and on an understanding of the differences between platform and ecological strategies.

In summary, innovation is the most central activity we see in business in the digital economy. However, innovation does not happen consistently. According to Clayton M. Christensen, innovators are bound to suffer from an "innovator's dilemma" [29]. Simply put, if an innovative organization acts in order to achieve

Z. Liu, *Principles of Digital Economics*, Contributions to Economics,
https://doi.org/10.1007/978-981-16-9020-4_9

short-term results, it usually creates a condition of long-term failure, which refers to the short-term positive value and long-term negative effects of corporate genetics.

In this chapter, two issues will be explored, including how the "innovator's dilemma" is broken by the enterprise of the future (the networked organization) and an analysis of the evolution of dualistic organizational structures and management logic. In line with these themes, the following three chapters will be discussed, which includes the dual structure of networked organizations, network platforms and ecological construction and innovation strategies for networked organizations.

Dual Structure of Networked Organizations

First of all, the two-tier structure of the network organization will be discussed, which is the most significant difference between network organizations and market and business organizations. The two-tier ecological structure of the platform and the user together constitutes the network organization. That is, the network organization belongs to the platform that generates the products and services, while the users become an important part of the ecology as the platform expands. This particular structure allows us to see an adjacency-type contract being formed by the future enterprise (network organization) under the blockchain through peer-to-peer agreements, replacing the atomic-type contracts of the industrial economy. At the same time, a dual structure of flatness and zero transaction costs emerges. The economics of this structure will be discussed in more depth later on. Here, the benefits of such a two-tier network structure are first and foremost discussed.

In fact, the theory of networked organization has already been very well developed in the Internet era. The "network effect" is its main theory. Here, the previous theories of network effects will be briefly introduced. On the one hand, the theory of network organization in the digital economy is still based on the development of the network organization of the Internet, so it has relevant economic properties. On the other hand, if these properties can be understood, they can also lead to a deeper understanding of the network economy effects in the digital economy.

The supply-side scale effect and the embodiment of the demand-side scale effect are the main components of the discussion of the network effect in the Internet era, which is the greatest value in the Internet economy.

The supply-side scale effect refers to the emergence of a firm's cost advantage due to the growth of scale and the reduction of unit costs, which leads to the formation of a monopoly effect. The supply-side scale effect brings not only huge investment and resource thresholds, but also the growth of the learning curve in the accumulation of scale, which allows efficiency and cost advantages to be realized. The demand-side scale effect refers to the network externality of interest, where each user who joins the network adds value to the network itself while enjoying the value of use. As the size of the user base increases, the network becomes more attractive to non-participating users. This is the essence of the social network business model, which does not rely on the product itself to create user value.

With the development of the digital economy, new qualities have emerged in the theory of network organization in the digital economy era, which include the following three main aspects.

Firstly, the Chicago School theory of the exclusive and exclusive use of capital was broken by the organization of networks as the main resource allocation for firms. As a result, the theoretical basis for the idea that monopolies harm markets is also broken. In the digital economy, most services and products are in fact monopolized. This monopoly phenomenon is a manifestation of the ecological scalability and efficiency of network organizations. And such platforms or ecologies are split up through anti-monopoly mechanisms in a way that is actually very much not free economy. But in Europe and the United States, where antitrust prosecutions are constantly being initiated, its ecological qualities on behalf of network organizations are not seen. Networked organizations have both a corporate and a market nature, so the issue of efficiency is discussed in the context of networked organizations, which refers respectively to platform efficiency (efficiency within the organization due to specialized division of labour) and eco-efficiency (efficiency in the diversification of the ecology formed by the expansion of the organization outside).

Secondly, the theory about the division of market structure by industry is broken by the dual structure of network organization. In industrial economics, only monopolistic behaviour based on industrial structure is taken into account, but not monopolistic behaviour based on the ecology of the firm. So in the Internet industry, "giant" organizations have emerged. They have not only built up an absolute core position in their own ecology, but have also spread their influence to other areas through acquisitions, mergers and acquisitions and their own innovations, such as the competition between Alibaba and Tencent in China, and between Google and Facebook abroad. The competition is generally at the ecological level rather than the industrial level. This is due to the dual structure of network organizations. Corporate qualities are explained by the platform, while industrial qualities are embodied by the ecology. Although the two are completely different concepts in traditional economics, they are concepts of the same property in the context of the ecology of network organizations.

Finally, the boundaries defined by property rights are relaxed by network organizational structures in order to shift the core of property rights from ownership to use, thus making the sharing economy an important proposition in the digital economy. In traditional organizational structures, the proprietary nature of a firm's capital leads to the fixing of the firm's boundaries. The boundaries of the products and services offered by enterprises are also limited, so it is difficult to circulate social capital resources between enterprises. In the digital economy, on the other hand, network organizations have no fixed boundaries. In addition to internal platforms and ecological services, network organizations can provide products and services to the ecology with the right of use as the core, so that the efficiency of the whole ecology is enhanced. This is why the property rights theory of dominance is broken down by network organizations. At the same time, a system of rights and interests centred on the right to use is constructed by it.

Of course, although network organizations have not only platform characteristics but also ecological characteristics, the more centralized network organizations are called "platform network organizations", which correspond to the mass customization approach to production. The other type of organization, which is more clearly "decentralized", is called an "ecological network organization", which corresponds to a personalized production approach. Both of these have the characteristics of network organizations, but they have a different bias, which makes them serve different industries and different ecologies. This needs to be noted. The change in production relations and the distribution of rights is being driven by the blockchain economy through distributed ledger technology in order to redistribute the interests of the entire networked organization of businesses, so that the entire business ecology is stimulated to become a self-organizing network of "technology contract-based stakeholders". This is why we believe that the sharing economy will become the mainstay of the digital economy, and that it implies the imminent emergence of a new networked organizational ecology.

Network Platforms and Ecological Construction

After the dual structure has been discussed, topics related to the ecology and empowerment of network organizations will be discussed. As we have discussed before, the economy in the digital economy can be understood as an organization that builds an ecology with a platform as the main body, which refers to the implementation of mass customized functions. The platform serves the application companies in the ecology through the provision of infrastructure that can be shared, which is essentially a homogeneous asset provided by the platform. The ecology then plays a differentiated value-added role, with homogeneous assets being transformed into heterogeneous services, and a diversity of innovative services being provided on a specialized basis.

From this perspective, the return to the classical economy and the re-sublimation of the industrial economy is the focus of the future digital economy, which refers to the return to the classical economy, i.e. the return to a value-centred economic form. The retention of the capabilities of the industrial economy at the infrastructure level is a sublimation of the industrial economy, which gives a basis for the effectiveness of a diversified value economy. The role of the digital economy in such a structure is to be able to achieve several goals that could not be achieved in the industrial economy.

Firstly, the fixed investments of the industrial economy are replaced by a shared asset approach in a networked-based organizational ecology. This is not only the basis for a future blockchain network-based sharing economy being developed, but it is also the work that the Internet era failed to accomplish. The elements of finance and price cannot be provided by the information network of the internet and intermediaries cannot be completely removed by it through decentralization. So the sharing economy stops at shared accommodation and shared travel. However, this economy of sharing is being expanded by the blockchain network to include almost all digitized services and products.

Secondly, resources and factors are redistributed through the operation of a virtual service and digital ecology with the service economy and experience economy at its core, in order to achieve another distribution of wealth. This time it is a value-centred, rather than a capital-increasing, form of economy. This will lead to a more equitable and just society and economy as a whole. The traditional logic of equity-based distribution of business organizations is reshaped. All users, providers of products, managers and investors will receive a share of the proceeds within the same ecology. Instead of only the returns of the investors being taken into account, the basic landscape and business model of the digital economy will be significantly changed.

Finally, a new collaborative mechanism and economic evolutionary ecology is being constructed based on the networked organizational ecology. The reason why networked organizations are used as innovation agents in the digital economy is that the capacity for continuous innovation can be acquired through continuous evolution. At the same time, new "symbiotic organizational relationships" are created based on a logic of cooperation between organisms, so that a new organizational form is formed based on the creation of value for users and efficient collaboration in the ecology. In this symbiotic network, all the ecological participants are mutual subjects, resources are shared, value is created and benefits are shared, and the evolution that could not be accomplished by the original atomized business organization is accomplished by a consensus-based technology contract.

In the light of the above conclusions, the governance of such a networked symbiotic organization will be discussed. In other words, the governance of platforms and ecologies will be analyzed and the following is its basic logic.

Firstly, the platform theory in the Internet era focuses on the bilateral market theory. In other words, platforms are considered to be intermediaries between bilateral stakeholders of supply and demand, with the needs of both parties being met by the platform's aggregation. Under this framework, the transactional attributes of the platform were the focus of attention, while the core function of the platform of providing infrastructure and the structure of how the platform forms a new business ecology with the participants was ignored. Instead, we believe that under the progressive development of blockchain technology in the network economy ecology, platforms have to be re-understood. Platforms should be understood as a whole ecological unit with multiple participants. At the same time, the development of the platform itself should not be seen in isolation. A property relationship of "use rather than possession" is provided by the platform.

In addition to the logic of bilateral markets, the 'duality' of business and market that exists within the platform itself needs to be understood. As we discussed earlier, firms and markets are analyzed by Ronald H. Coase as the two main forms of resource allocation. Coase analyzed them as the two main forms of resource allocation. Platforms, on the other hand, have both firm and market properties, with network-based organizations forming the main concrete form of platforms.

Information platforms are the main manifestation of platforms in the Internet era. In contrast, with blockchain technology, decentralized distributed platforms are formed with communities and communities as the main focus. In either form, they have dual attributes at the same time. On the one hand, platforms have economic elements such as free assets and labour, which are aligned with corporate organizations. On the other hand, platforms are marketplaces that do not sell products directly but offer a particular service or product. Supply and demand are matched by it. This is consistent with the role of the market. Therefore, platforms are the attributes of firms that compete in the marketplace, and their greatest quality is the "externality of networks across borders". Simply put, the value of a platform increases as the number and scale of bilateral users increases. Platforms are a marketplace for both supply and demand, and are characterized by a "monopoly of market power". This means that bilateral transactions are facilitated through the provision of high value products and experiences. In summary, the key lens through which we understand the inner logic of platform attributes is the "dual attribute".

The phenomenon of incremental payoffs is not uncontrolled and its worth drawing attention to. Incremental payoffs for platforms fail when the value of the infrastructure provided by the platform is lower than the value of the infrastructure service facilities created by these companies themselves. So the idiosyncrasies of the platform need to be considered when it comes to its governance. One of the most important tasks is how to provide the best optimal social capital for the ecosystem and the completion of the empowerment of each node of the ecosystem. A platform can be understood as a product of the 'economies of scope', where social capital based on shared value is consistently, consistently and comprehensively provided within a network. Therefore, the key to platform governance is how to empower and how to add value to the assets in the process of empowering the ecosystem.

So how can platforms empower the ecology? This needs to be discussed in terms of the communities within the platform and the communities that the platform empowers externally, which can be considered from three specific perspectives.

The first is that for the platform itself, 'employee autonomy' needs to be inspired by a shared vision to achieve a culture of community innovation within the platform. Network organizations rely on shared values and goals to achieve innovation. After this, we want the employees of the internal ecosystem to form an autonomous organization. At the same time, three levels of evolution are accomplished, including a high level of team chemistry, innovation that guarantees speed and timeliness, and flexibility and adaptability within the organization driven by self-organizing communities. In a traditional business organization, the goal of the employee is to fulfil the tasks in the job description. In a networked organizational ecology, on the other hand, each employee is the fulcrum of the organization's strategy and they rely on a shared vision to achieve their mission. This self-organization is what the ecology within the platform needs to achieve.

Secondly, the platform needs to provide the infrastructure that should be used externally in order for the basic functions of the entire ecology to be safeguarded. It guarantees the storage and operation of information through hardware and the security and efficiency of information through software. It means that its products and services are exported to all parties in the ecology through big data and cloud services to provide infrastructure for the whole ecology. At the same time, the credit of the entire technical contract of the transaction is guaranteed by blockchain technology, so that the platform has the most complete infrastructure in the digital ecology.

Finally, a service and experience-based ecosystem needs to be created by platform companies. In the past, the "digital" part of the digital economy has always been emphasised. In fact, the essence of digitalisation is that after the infrastructure (fixed assets) has been provided, we need to consider the following core qualities of the digital economy, which refer to how the value generated by these assets can be serviced and experienced, i.e. the creation of value added based on personalized needs. In short, a digital ecosystem based on the premise that assets are not proprietary needs to be built by platform companies in order to enable the 'incremental payoff' of the economies of scope. This economy can only be highlighted in an economy that is digital, service-oriented and experiential.

The above is a basic discussion of the construction of network platforms and ecologies, especially the study of the management principles and governance mechanisms of symbiotic networked organizations. In the context of the digital economy, networked organizations, as distinct from the industrial economy, are built up by business organizations through platforms and ecologies to enable the digital economy's unique business model based on incremental returns and network effects to be realized. At the same time, a "symbiotic" collaborative relationship and management concept is formed in this network organization ecosystem. An understanding of the economics of this not only facilitates innovators in building their own network ecology of business organizations, but also facilitates the understanding of the qualities of the digital economy from the practical side of business.

Innovation Strategies for Networked Organizations

The dual structure of the network organization is discussed followed by the ecological strategy of the innovative organization. This issue is not limited to the new economy, its a paradox that has occurred continuously throughout the history of business innovation. Business history is full of companies (such as most of those described in Built to Last: Successful Habits of Visionary Companies [30]) that succeed and then fail, as well as innovate and then slacken. The fact that markets and technology are constantly changing in unpredictable and complex ways leads to the paradox described above: the successful innovation gene on which successful organizations are based leads to a situation of dynamic conservatism, where they actively protect their old competencies. Timely adjustments are therefore difficult to make in the face of new changes in the business environment and technological change. The innovator's dilemma means that the innovation cycle is affected by technological evolution, so that management styles and organizational structures are called upon to undergo a new evolution. At the same time, managers need to know how to avoid the "innovator's dilemma" (which refers to the pitfalls of success) and how to encourage innovation in their organizations while improving the efficiency of management. These are not only challenges for all companies, but they are also the starting point for our discussion of innovation management in networked organizations. In this chapter, the elaboration of an ecological strategy for networked organizations is discussed, in particular the innovation strategy for networked organizations in the digital economy.

The key elements of a networked organization's innovation strategy are discussed first, and they include a value proposition and an element of value. The intrinsic logic of the value internet had been mentioned in terms of its informatization, networking and the value of social capital. Here, the connotation of value will be understood at the level of eco-strategy, which refers to the value network as reflecting a certain value proposition of the network organization. The value proposition is used by the network organization to configure the value elements in order to create the basic structure of the ecology. The essence of the value proposition is in fact the basic elements of the consensus mechanism of the blockchain network we are discussing. Different value propositions are formed by different consensuses, which dominate different ways of allocating resources. The digital network is empowered by the traditional enterprise organization to upgrade its value proposition. Moreover, this is the essence of digital eco-strategies such as "Internet+", "eco-empowerment" and "digital transformation".

The economic resources that must be available to realize the value proposition are the value factors, which include productivity, products, services and other resources. The logic of ecology is based on the integration and allocation of value factors based on ecological resources, rather than the logic of corporate cooperation. The major difference is that users are also key components of the value factor and participate in the realization of the value proposition. Simply put, a platform is not only a value element, but also a specific ecology. The value proposition can be resourced in a cost effective manner through the platform model. In general, the different value elements of the ecology are finalized through the participating actors and their division of labour and distribution in order to enable the dynamics of the ecological configuration to be realised and thus their value proposition to be realized. This is the essence of an ecological strategy.

The relationship between open innovation and networked organizations needs to be looked at. Business organizations have been studying open innovation to improve the competitiveness of their companies for several years. However, the real value and impact of this research has emerged in the digital economy, especially since the Internet was developed. The main difference between open innovation theory and conventional innovation theory is that a company's competitiveness is enhanced through the use of external knowledge and the creation of a learning organization, and the company's own boundaries are transcended. And this openness is most evident in the context of the digital economy. It is almost futile for networked business organizations to attempt to lock up intellectual property within the company's value chain. At the same time, customer demand has shifted from being thing-oriented to event-oriented, meaning that people no longer focus on the specific product delivered but, rather, on the experience and service it brings. In this context, the value chain has to be diffused from vertical to horizontal, from a closed linear system to an open and complex ecosystem, and from competition to synergy and cooperation. This is not only the inherent logic of open innovation, but its also in line with a question we have discussed before, which refers to the problem that networked organizations in the digital economy need to solve in the complex economic ecology of the digital economy is how to achieve knowledge sharing and value co-creation.

The three basic stages of a networked organization's innovation strategy will then be discussed, which include digitalisation, platformisation and ecologisation. These three processes are inseparable. The transformation of a traditional business requires a change in the ecological strategy based on its value proposition, which consists of the following three stages: (1) it changes consumer habits and builds a digital strategy through digitalisation; (2) the formation of a digital business ecology is driven by the platform element; and (3) the reconfiguration of the value proposition and value elements is driven by the ecological strategy. This process is an interlocking rather than a linear one, which is noteworthy. Moreover, from the perspective of strategy studies, digitalisation has a higher connotation than ecological strategy, and ecological strategy is in turn higher than the category of platformisation. The basic logic of the digital economics we study is that the digital transformation of traditional companies is treated as the focus, rather than the traditional companies themselves being treated as the focus. Platformisation is the ecology of specific value elements and specific value configurations, and it is the most common ecological strategy in the Internet era. Ecologisation is a broader strategy of value allocation, which does not necessarily require the emergence of a platform role. Simply put, ecologies can be built directly by network organizations without the need for a platform.

The cognitive upgrading of strategies for networked organizations in the digital economy will be discussed at the end. Basically, platformisation and digitalisation are already known to most readers, so topics related to digital ecology will be discussed mainly. This is because there are many questions that need to be answered. The central goal of the network economy in the digital age is that value factors are configured through value propositions so that the reshaping of the business ecosystem is realized and an effective network organization and business ecology is formed with almost zero transaction costs. Furthermore, the nature of this digital business ecology will be interpreted in three ways.

1. Business ecology refers to the formation of multiple related or unrelated business modules based on the construction of a core business model, and the formation of a complex network around these business modules. The economy has always been built through evolution and iteration, rather than through planning. Therefore, the most important thing for innovators to focus on is the evolution of the ecosystem driven by different value propositions. At the same time, the design of the digital economy ecosystem has to be considered first and foremost. The

business ecosystem only needs to be responsible for setting the rules, not planning each step.
2. In the process of ecological evolution, it is divided into two main basic logics, which include ecological exploitation and ecological evolution. The former is the process by which a basic ecology is constructed, while the latter is the process by which the boundaries of an existing ecology are expanded. Through the multiple cycles of ecological creation and expansion, innovators drive the entire business ecology to iterate under established rules. In this process, the following two basic logics are to be attended to.

Firstly, markets are uncertain. Therefore, dynamic strategic thinking has to be established, which refers to the ability to constantly innovate and develop flexible strategies in complex markets.

Secondly, the ecological strategy is divided into an internal orientation and an external orientation. The network organization must not only drive the implementation of the internal organization's value proposition for products and services, but also the continuous entry of external organizations into the ecology so that network effects can be realized as quickly as possible.

Thirdly, the strategy has shifted from competition being to integrating resource allocation to cope with the changing external market environment, which is the core of ecological competition strategy. The business models and product portfolios emphasised in the Internet era have been gradually phased out, and the competition of the future is between business ecologies. Ecological evolution and the ability to adapt to the environment are at the heart of competition.

In summary, innovators need to determine which elements can enter the ecology in the process of ecological evolution. At the same time, they should give different productive relations and incentives to different elements according to them, and emphasise their access rather than ownership. This is the key to a shift in our strategic thinking. Based on the above understanding, an innovation strategy for networked organizations can be established, which can also be called an open innovation strategy in the digital economy. Its essence is the realization of the value proposition in the ecology being driven through the logic of digitization, platformisation and ecologisation, and the configuration of value elements in the ecology.

Chapter 10
Property Rights and Equilibrium Theory in Digital Economics

In this chapter, the property rights and equilibrium theories of digital economics are discussed. Here are three reasons why this chapter is being explored.

Firstly, the importance of the sharing economy in the blockchain economy has already been discussed. The right to use is the central part of the sharing economy, and the right to use leads to a discussion about "rent". This part of the discussion concerns the logic of the evolution of property rights theory from ownership to use rights, which is intrinsic to the development of the sharing economy.

Secondly, the issue of property rights in the digital economy needs to be discussed initially, mainly in terms of the relationship between rights and values. The separation of use and ownership is the essence of the sharing economy. How is the impact of the digital economy and the technology represented by blockchain on this area to be understood?

Finally, after the basic property rights issues have been discussed, the basic logic of the governance system of the blockchain will be discussed on this basis in order to allow practical solutions and logic of property rights governance to be proposed. The governance system in the digital economy is increasingly using many of the logic of policy making in economics, and its one of the biggest differences between the corporate governance system in the digital economy and the traditional governance era.

Equilibrium Theory of Digital Economics

We have discussed that classical economics was replaced by neoclassical economics and that an important reason for this was that the utility-based theory of marginal productivity was provided by neoclassical economics, which refers to the general equilibrium theory. The lack of equilibrium theory led to a lack of generality in the value theory of production costs in classical economics, as well as the inability of many economic phenomena to be discussed in the context of classical economics. In

Z. Liu, *Principles of Digital Economics*, Contributions to Economics,
https://doi.org/10.1007/978-981-16-9020-4_10

neoclassical economics, on the other hand, the factors of production are proposed to have value, the extent of which is determined by the marginal utility obtained from the consumption of the final product produced by these factors, and not by the cost of the product itself, as understood in classical economics. The simplest and most classical examples are as follows: diamonds are more expensive but have a lower utility; water has a lower price but a higher utility. This phenomenon cannot be explained by classical economics. They understood this phenomenon in terms of the overall utility that diamonds and water bring to the consumer, whereas it is explained by neoclassical economics through marginal utility.

The equilibrium theory of digital economics is therefore to be built up. On the one hand, the deficiencies of the theory of classical economics need to be remedied. On the other hand, the discussion of utility in neoclassical economics has to be refuted. So two basic frameworks are to be provided simultaneously. Firstly, the ideas of neoclassical economics concerning equilibrium theory have to be assimilated in order to develop a theoretical framework suitable for digital economics. Here, the model of increasing payoffs constituted by technology-based endogenous growth is to be discussed in particular, with its reference to the analysis of the endogenous growth model proposed by the Nobel Prize-winning economist Romer. As well as the inner logic of the growth of the digital economy is understood through the differences between his theory and the neoclassical economic growth theory proposed by the economist's Solow. Secondly, Zhang Wuchang's discussion of the new institutional economic theory needs to be assimilated, particularly the theory of rent in contract theory. The framework of the usage-centric sharing economy is understood through the concept of rent and the analysis of heterogeneous capital.

Firstly, the ideas of the equilibrium theory of neoclassical economics need to be discussed. At its core, general equilibrium theory means that the relationship between supply, demand and price is explored within the framework of multiple interacting markets in the economy as a whole, in order to build a complete market model in which consumers, markets and firms all achieve maximum rationality. Neoclassical economics was developed on the basis of general equilibrium theory, and the economist Robert Solow proposed a model in the 1960s that was based on the theory of general equilibrium. The neoclassical theory of economic growth, developed by Robert Solow in the 1960s, is a truly modern set of growth theories based on general equilibrium theory. This theory won him the Nobel Prize in Economics in 1987. To this day, this model remains an integral part of economic growth theory.

The growth model constructed by Solow is simple. In this model, the accumulation of capital can be achieved by the economy through the generation of a savings component. This accumulated capital has two purposes. On the one hand, it is used for the "broadening" of capital, which is equivalent to providing capital for the new population. On the other hand, it is used to facilitate the "deepening" of capital, which means that the per capita capital stock in the economy is raised. The marginal output of capital is decreasing. Thus, the economy reaches an equilibrium as capital accumulates, meaning that when capital deepening is exactly equal to zero, all new savings are used for capital expansion. In equilibrium, both the capital per capita and the corresponding output per capita in the economy will remain fixed. At the same time, the standard of living of people in the economy will also remain constant.

So what determines the equilibrium capital stock per capita and output per capita? In the Solow model, it is determined by several factors, including the rate of population growth, the savings rate and the level of technology. When the population growth rate is higher, more capital is needed for "diffusion". The equilibrium capital stock is thus reduced. At the same time, higher savings rates lead to higher capital accumulation, which leads to an increase in the equilibrium capital stock. In addition, a higher level of technology can produce more output for the same capital input, leading to more accumulation, which also leads to a higher equilibrium capital stock.

Based on the predictions of the Solow model, when the rate of population growth, the rate of savings and the level of technology are fixed, the level of capital per capita and output per capita will converge over time towards the equilibrium level, regardless of the starting point of the economy. It follows empirically that the population growth rate and savings rate of an economy will generally remain constant over a long period of time. So ultimately there is only one variable that determines an economy's equilibrium level of development, and that is technology, which can be more accurately referred to as Total Factor Productivity (abbreviated as TFP in the following). However, the Solow model is not perfect. On the one hand, although the importance of TFP in the growth process is recognized by the Solow model, it is not clear what factors determine the Solow model. In fact, in the Solow model, TFP is a completely exogenous variable. On the other hand, some empirical findings also conflict with the predictions of the Solow model. For example, the prediction based on the Solow model is that the economic levels of countries should converge towards the steady state and that incomes should converge between poor and rich countries. In reality, however, per capita income has been growing consistently in many rich countries, and convergence between poor and rich countries does not seem to be evident.

In the light of the above paradoxical dilemmas of theory and reality, the Solow model was reflected upon and explained by the economist Romer. In 1986, he published his paper Increasing Returns and Long-Run Growth [3], in which the concept of "incremental payoffs of scale" was introduced to explain sustained growth. He argued that the emergence of knowledge as an element of incremental payoffs led to the emergence of incremental payoffs of scale by emphasising its positive externalities. In this model, he proposes a four-factor theory of growth, which includes not only capital and labour in neoclassical economics, but also human capital (years of education are measured and knowledge is emphasised) and new ideas (patents are measured and innovation is emphasised). Romer assumes that knowledge has the following properties, which include positive externalities, increasing returns in the production of output and decreasing returns in the production of new knowledge. The competitive equilibrium growth production model is made up of these three factors together and concludes that output will grow positively even in an economy with zero population growth. This is Romer's greatest contribution to growth theory.

Here, it is the influence of Romer's 1986 article on the equilibrium and growth theories of digital economics that needs to be focused on, which can be broadly summarized in three parts.

First, knowledge, a factor of production with externality properties, is emphasised to lead to an aggregate output function with increasing returns so that economic growth can be sustained over time or even grow faster and faster. The model shows that the market resulting from perfect competition is not optimal and that monopolistic competition is necessary. Monopolistic competition leads to sustained economic growth, provided that property rights are protected. So in digital economics, the growth of factors such as knowledge and innovation is also focused on, rather than the growth of labour and capital.

Second, how can increasing returns and markets be unified by perfect competition in a general equilibrium model? This needs to be explored. A dynamic equilibrium theory of increasing returns is provided by an extension of the general equilibrium theory. The equilibrium theory of digital economics is also based on such an extension of general equilibrium theory, where the idea of a dynamic equilibrium theory of increasing returns is provided and a dynamic equilibrium is considered to be formed by the increasing returns of new factors (which mainly refer to the factors of innovation and knowledge) and the decreasing returns of other factors (which refer to labour and capital) so that the dynamic equilibrium of the digital economy is achieved.

Finally, the competitive mechanisms of the market in this scenario are articulated to lead to sub-optimal outcomes. Therefore the market should be intervened in by government through a combination of taxes and subsidies in order to endogenise these externalities and thus bring them back to the social optimum. This would also provide a theoretical basis for the theory of innovation in digital economics, and in particular for the support of innovation by government policy. Markets that are completely free from government intervention and free to compete are usually ineffective. In suitable scenarios, government must intervene to support the development of industry and innovation in order for the industry to develop positively.

The concept of rent in economics will then be discussed. Because this concept is particularly important in the theory of numerical economics, it is to be introduced in detail.

First, the concept of rent is analyzed in terms of the evolution of economic history. Rent is not only the income of a factor of production, but also the value of an asset. It first referred to the rent of land. Agriculture was the primary concern of the classical economists in the early days and the most important factor of production in agriculture was land rent. The classical economists argued that labour produces value, so that the collection of rent by landlords could not be explained. This is what led neoclassical economics to abandon the historical cost theory and replace it with a new concept of opportunity cost. Marshall's study of land rent, for example, revealed that the income of all factors of production was their own rent, and that "land rent" was merely a special case of rent. In other words, the allocation of resources under the constraints of selfishness and scarcity is the basis of neoclassical economic research, and the prices of goods are determined by both of them. Thus, the income of the producer of the good is constituted by the price. This income is the rent.

Secondly, rent is analyzed in terms of changes in income. If income changes but supply does not change with it, a rent is created. In traditional economics, profit is usually used to discuss income minus opportunity cost. In fact, the concept is more appropriately interpreted in terms of rent, i.e. rent is equal to the value of real income minus opportunity costs. When people make choices, they consider the maximization of rent. If the first perspective is that rent is understood as the income of a factor of production or the value of an asset, then the second perspective is that which allows the outcome of an individual's choices to be explored in a more crude way, i.e. the change in rent.

Finally, rent is discussed in terms of the basic assumptions of economics. The selfish person will seek to maximize rent, its a more appropriate dimension than maximizing income as well as minimizing costs. In the most extreme case, rent is equal to costs. Income, on the other hand, is expressed as cost itself, except that its expression is divided into visible and invisible costs, but there is no difference in its essence. In other words, the selfish nature of man and his competitive behaviour dictate that rent is in the extreme both a cost and an income, and that income is in fact both a cost. There is no income that is not a cost. This is an important perspective in our understanding of rent.

After the concept of rent in economics is understood, the understanding of value theory in the digital economy under the concept of rent needs to be complemented. Thinking is switched from a price perspective in neoclassical economics to a value perspective. Simply put, rent is understood in the context of a heterogeneous dimension of choice, rather than in the context of homogeneous competition. We have discussed that in digital economics, the satisfaction of differentiated needs of different people is emphasised. If such economic effects are to be measured, they need to be understood in terms of the dimension of heterogeneous rent. This is how the economic model of the digital economy, where diversity and specialization co-exist, can be understood.

Finally, the equilibrium theory of digital economics will be summarized. The existence of rent allows us to see the existence of two types of rent in a differentiated economic model, understood from two perspectives. On the one hand, from the point of view of resource allocation, the differentiation between goods is rent. In short, rent is heterogeneous profits. On the other hand, from the point of view of the distribution of benefits, the differentiation of rights between people is rent. In short, it is the cost of use. In the digital economy, the entire ecosystem is run by the platform through the acquisition of rent, which are divided into two types of rent: one type of rent is a fee for the right of use of the end consumer, while the other is a fee for the infrastructure used by the businesses within the ecosystem. Both types of rents are brought about by heterogeneous utility. In this process, the equilibrium state of the digital economy is created.

Here, the concept of rent dissipation needs to be added. This concept is very well known for the reason that in "The Structure of Contracts and the Theory of Non-Exclusive Resources", Zhang Wuchang derived and analyzed the theory of rent dissipation for non-exclusive resources from previous studies of marine fisheries by economists. And indeed, the essence of the concept is that the perspective of neoclassical economics is being used to explore the laws of value in classical economics. That is, the phenomenon of price fluctuations around value is explained. If the concept of rent is used to explain this phenomenon, it refers to the phenomenon of various forces playing against each other moving around the point of resource allocation under Pareto optimality. From the perspective of homogeneous rent, Zhang Wuchang deduces that homogeneous rent dissipates as a result of homogeneous perfect competition. Thus, the dissipation of rent needs to be reduced by the definition of private property rights. And the right of ownership needs to be absolutely monopolized. Furthermore, from the perspective of heterogeneous rent, it follows that dissipation of rent triggers homogeneous competition and vice versa heterogeneous perfect competition. Defining property rights also requires transaction costs, so the perspective of heterogeneous competition underpins the discussion of the relativisation of ownership in the digital economy.

Once the principles of rent value and rent value dissipation are understood, the equilibrium theory of digital economics should be understood based on the concept of rent value. In a digital economy, rent value corresponds to the difference between the equilibrium price of monopolistic competition and perfect competition. That is, if competition is the difference between the returns from monopoly and the returns from a perfectly competitive market, the rent value of monopoly returns will be formed. This is the theory of equilibrium in the digital economy. In fact, all resource allocation moves towards monopoly and perfect marketization. Eventually, a complete monopoly will emerge as a result of elements such as technological innovation. The winner then receives all the gains and the innovator then opens up new markets through innovations that recombine technological elements. In the digital economy, rent corresponds to the rights of users, and its the reason for the sharing as the core model. At the same time, rent corresponds to a monopoly barrier for organizations to enter a certain field. All intellectual capital, when differentiated and personalized (especially organizers and workers with the capacity to innovate), must be scarce, thus creating a heterogeneous competitive environment that relies on the innovative capacity of entrepreneurs. The goal of this competition is to seek a monopolistic equilibrium in a given field.

The Property Rights Theory of Digital Economics

After the equilibrium theory of digital economics and the problems associated with rent have been discussed, the theory of property rights will be formally discussed. This issue is not only the focus of research in the new institutional economics, but it is also one of the most talked about issues in the theory of digital economics. We have mentioned that the new institutional economics is the essence of contractual economics. Therefore, after the traditional theory of new institutional economics and the definition of the property rights issue have been introduced, the theory of property rights in the concept of digital economics will be explored.

The discussion of property rights had to involve transaction costs [31], and the discussion of transaction costs needed to involve Coase's theorem. Three different versions and connotations are offered by Coase's theorem, with textbooks generally being the third version. We present them all here for you.

The first version of "the definition of rights as a necessary precondition for market transactions" means that all assets are considered to have private property rights, including the right to use, the right to income and the right to transfer. This does not include ownership rights. This is because once the first three rights are present, ownership is in fact present. The sale of property in the country is not in fact a sale of ownership rights, but a combination of the first three rights.

The second version of this is that "the use of assets under a market transaction results in the highest asset value when there are clearly defined property rights". Of course, this conclusion is not entirely valid in fact. In economics, the "selfishness" qualification assumes that all behaviour is already optimal. So the true value of assets is not affected by whether or not there is a clear definition of property rights. At the same time, both the definition of rights and the construction of markets inherently require transaction costs. So, given the existing limitations, all assets should be at their highest (Pareto Optimality or equilibrium optimal). Thus, this version actually expresses a correct nonsense and the definition of rights costs in Coase's theorem is exposed by it.

The last version of the statement that "rights, as long as they are clearly defined as private property rights, end up the same regardless of who they are initially defined to, and under the condition that transaction costs are zero, the outcome of market transactions is the same". It is the most widely known element. In other words, markets and other institutions have transaction costs. If the institutions are consistent in their effects, random unpredictable market behaviour arises. Thus, there exists the condition that transaction costs are zero to derive the conclusion that no market system is needed. This premise has been changed by Chang to "the relative price of the market remains constant" in order to make the conclusion more reasonable.

Once the theory of property rights in the new institutional economics is understood, the theory of property rights in digital economics will be discussed. It is the equivalent of the "use rather than possession" property rights relationship we have discussed, which refers to the sharing economy. In the old economy, there was a symmetry between use and ownership; in the digital economy they are asymmetrical. In industrialized thinking, ownership and dominance are put together in a way that is paramount. The famous German jurist and founder of the historical school, Friedrich Carl von Savigny mentions in "The theory of possession" that "possession is not only a right but also a fact". In other words, the duality of possession is the same as the duality of goods. Civil law possession and natural possession are discussed by Savigny from the perspective of juridical research. The former is possession in the legal sense, which concerns the relationship between man and man, and which is considered primarily in the dimension of value. The latter is possession in the juridical sense, which concerns the relationship between man and thing, and which leads to the value of use. From this perspective it follows that the former is a natural right, while the latter is a social right. In other words, natural rights correspond to use values, while social rights correspond to values. This derives from social relations to the question of ownership. This has not happened in the long history of Western jurisprudence. Before the industrial revolution, "dominance" was emphasised over "ownership". After the industrial revolution, the concept was stolen by the bourgeoisie as "ownership". The adoption of a "use rather than ownership" approach to property rights can therefore be seen as a radically reform.

Depending on the property relations of access rights, changes in the theory of digital economics based on network effects related to property rights can then be discussed. The mechanism of cooperation under the prisoner's dilemma and the law of scale arising from network effects are the key concepts here. At the same time, it is the basis for our understanding of the underlying logic of the economic effects of collaboration in networks. Property relations are centred on use, which leads to each individual being more involved in the creation of value, rather than thinking about their own value in terms of the outcome of income distribution. The basic ideas of the following two concepts will then be presented separately.

The mechanism of cooperation based on the prisoner's dilemma suggests that the creation of social capital is the basis for cooperation in networked organizations. We have discussed that social capital is all the elements that exist within a social network that facilitate cooperation in the prisoner's dilemma. The process by which these elements are created is based on collaboration. Because it is based on a general process of cooperation rather than a competitive game between the collaborators in the network, rational choice can produce effective collaboration under multiple iterations of the game, thus allowing individuals to contribute at the expense of their personal interests in order to meet the long-term development needs of the organization. This is similar to the different divisions of labour in a beehive. The ability of the colony to survive is preserved by the bees through the sacrifice of self. In the digital economy, on the other hand, network organizations are built on consensus. The prisoner's dilemma is disengaged through different collaborative divisions of labour. This collaborative and symbiotic relationship is premised on the role of "imaginary communities" formed through value networks by the "trust machine" of digital technology. On the one hand, the contribution and value of behavioral economics in this context needs to be seen to be strictly enforced through technical contracts that enable consensus. On the other hand, the ethical foundations of this need to be seen. The foundations of networked organisations are ethical, able to call on each individual to sacrifice individual short-term interests for the long-term values of the organization.

In general, property rights are discussed in terms of resource allocation in neoclassical economics, which refers to the interrelationship of material forces in the economy. On the other hand, the interaction between social forces is discussed in terms of the interaction of interests in classical economics. The modern system of property rights has been built around ownership. In fact, we believe that the logic of contract theory, in which the right to use is placed at the heart of property relations as a central logic, is a very important change. That is to say, contract-based contracting, the identification of shared ownership of assets and the shedding of exclusive property relations is a theory of property rights based on cooperation and benefit sharing. The cooperative mechanisms of networked organizations and the scale effects that occur are based on a change in property rights theory that spreads to all digital assets and ecologies and is not limited to the scope of the information internet.

Blockchain Networks and Governance Mechanisms

After the issue of property rights in the digital economy is discussed, the inner logic of blockchain networks and governance mechanisms will be discussed, which belongs to the important area of applied digital economics property rights theory.

A governance mechanism based on a technology contract was created through a technology contract whose is the biggest impact blockchain technology has had since its inception, rather than the creation of a form of virtual currency. A crypto-pass-through economic system worth hundreds of billions of dollars was spawned by Zhong Bencong in just a few pages of a white paper. It brought forth a network of computers that is tens of thousands of times larger than the top 500 supercomputers. At the same time, it encompasses a diverse digital ecosystem of millions of developers, users and corporate organizations. In terms of social governance, this is undoubtedly the first time in the history of human civilization that self-organization has taken place without a clear central leadership. At the same time, it demonstrates the power of blockchain networks as the economic agents of the future. Huge and complex business ecosystems can be constructed through a completely self-organizing ecology.

The underlying logic of blockchain's governance mechanisms and its application to the theory of property rights in the digital economy therefore needs to be examined in detail.

Firstly, the basic logic of blockchain networks is discussed. The evolution and variation of blockchain networks is facilitated through incentive and coordination mechanisms to enable the formation of dynamically balanced business systems. At present, the technology of blockchain networks is still at a very early stage, so the core issues of blockchain governance mechanisms are the design of the underlying incentives and the establishment of coordination bodies. The former is achieved by solving the byzantine generals problem and designing a pass-through economic system, while the latter is achieved by setting up a community voting mechanism and reaching on-chain consensus. At present, they are both at a very early stage. Here, incentives and coordination mechanisms will be briefly analyzed.

The coordination mechanism means that once the technical contract is established, the network is ensured to evolve iteratively in the right direction. At the same time, coordination between chain and off-chain values is ensured so that the digital assets in the blockchain network correspond to the actual products and services. Currently, coordination mechanisms in the blockchain community are coordinated in a very geeky way. For example, coordination of the network development process is achieved by Bitcoin and Ether through communities as well as emails, etc. geeks' passion for the technology and trust in the community leaders is relied upon by such mechanisms. We hope that the coordination mechanism of the future will be a complete mechanism based on chain-based governance, complemented by off-chain ecology, so that much of the governance logic of the reality of economics can be proliferated and innovated by means of technology.

The wave of blockchain's empowerment is likely to start with the niche economy, rather than the disruptive mass economy, which is worth making note of. It offers a flexible and efficient ecology that enables niche economies to be recognized based on personalized value needs. And the pass-through economy ecology has emerged based on niche economies and personalized value services.

Based on the above reading, the pass-through economy and network effects can be linked. We have discussed the function and role of the network effects of the internet. Blockchain networks, on the other hand, possess a more complex value. The most important manifestation of the network effect of the internet is the model of monopolistic competition that drives it. The realization of win-win ecological value is driven by decentralized and driven blockchain networks. Products and services can be freely chosen by each individual, who does not have to worry about the barriers to transfer that come with monopolistic behaviour in the internet age. For example, it is difficult for people who used Facebook in the Internet era to stop using it, because they can hardly persuade themselves to run their friendship circles from scratch and sink their relationships and content in the new social ecosystem. However, the state of all applications and of each node in the present moment can be perfectly replicated simply by means of a fork in a blockchain network. At the same time, an almost perfectly consistent ecology of network applications can be created simply by the technical contract of a consensus mechanism. This is where the value of the blockchain network effect lies.

Finally, future trends in the governance mechanisms of blockchain networks will be discussed. In the future, one of the most central tasks of the entire business ecology of the digital economy will be blockchain governance. One needs to build the ecology of the blockchain network in an on-chain and off-chain manner to ensure the validity and legitimacy of its business ecosystem, so that each participant can have loyalty under a unified consensus. At the same time, the realization of network effects can be driven to attract more members to join. This is what we see as the greatest value of blockchain network governance. On the other hand, the governance mechanisms of blockchain networks are currently very imperfect. Therefore, in the process of continuous innovation, people have lost trust in the network and the community due to the erratic nature of the mechanism, which has led to the collapse of the blockchain network. This case is often seen in many well-known blockchain networks. At the same time, this is the price that has to be paid for the growth of the blockchain business ecosystem.

A trend we have mentioned is argued by the institutional arrangements of the blockchain economy, which refers to the original theories of economics, political science, etc. being used in macro-thinking, business management and institutional design in order for the landscape of business management theories to be expanded, as well as macro-theories being able to be applied to a smaller granularity of the practical process. For example, in the process of setting up a blockchain governance system, the founders of the blockchain network, like the pioneers of the US Constitution, have given a positive and iterative mechanism that can continue to develop and evolve to adapt to the external environment, taking into account the balance of interests of different parties and the long-term value of the entire ecosystem.

Part III
The Capital Theory of the Digital Economy

Chapter 11
Monetary Theory of Digital Economy

The next chapter will focus on monetary theory, which is the most important module in the second chapter. Why is it important? Here are three reasons.

Firstly, the main vehicle of the modern business community is monetary credit. Therefore, an understanding of the nature of money is essential to the understanding of the logic of commerce. If the statement that commodity credit and the functional functions of central banks cannot be replaced by the algorithmic credit of virtual money needs to be understood, then the origin and nature of money should be understood first. The logic of privately issued money does not hold for the foreseeable future. At the same time, the complete replacement of legal tender by privately issued money will not occur.

Secondly, economic growth and development are directly related to the understanding of monetary theory. The most important mission and duty of economists is to explore the theoretical foundations of inflation stabilization and to find good solutions to combat inflation. This presupposes that the monetary theories and solutions of traditional and Austrian economics need to be understood. When "stablecoins" were introduced in the blockchain space, many innovators were confused about the concept of money. They mistook stablecoins for "stable money". And this innovation in the digital economy has to be understood.

Finally, when blockchain technology was widely developed, the theory of cryptocurrency represented by the token economy was widely accepted. The discussion of the monetary nature of token is crucial. The theory we want to give is important for how the concept of token can be "dematerialized" and transformed from a "virtual currency" to an object of value within a "niche economy". At the same time, this is key to the understanding of not only token in the blockchain, but also to the large-scale commercialization of the blockchain industry.

Based on the above, this chapter is divided into three parts, which include the monetary theory of traditional economics, the monetary theory of Austrian economics, and niche economies and token economy.

Z. Liu, *Principles of Digital Economics*, Contributions to Economics,
https://doi.org/10.1007/978-981-16-9020-4_11

Monetary Theory of Traditional Economics

The hallmark of modern civilization in today's world economic system is undoubtedly money and the financial system based on it, which has given the international trading system and the social order its most basic institution. The behaviour of each country and even of individuals is regulated by it on the basis of the economic system of money. At the same time, the greatest problem facing the global economy is the economic problem posed by the inflation of money. In this chapter, the monetary theory of the traditional economy and the economic problems posed by inflation are discussed in three main chapters: the evolution and function of money, Keynesianism and inflation, and the limitations of the traditional monetary theory.

Firstly, the evolution and function of money needs to be discussed. The nature and social function of money and the functioning of the monetary system needs to be understood through the learning of the evolution of institutions. In the early days of human society, the medium of exchange did not yet exist. In this society of direct exchange, transactions were very limited and uncertain. In the economic model of direct exchange, there was a direct demand for goods or services from one party to another. As a result, there is a large element of luck involved in trading. At the same time, there are limits to the scale of transactions and the complexity of the division of labour and collaboration between members of society. And as society continued to develop, a set of indirect trading practices began to emerge. It refers to the expansion of the scale and scope of trade through the commodity as a medium of exchange, so that the emergence of a universal medium of exchange (money) is facilitated, not only by the satisfaction of the non-monetary demand for utility as a consumer good or factor of production, but also by the emergence of a monetary demand based on the exchange value generated as a medium of exchange. In other words, money has both a use value and a commodity value. Economic goods with these two attributes are defined as money. It follows from the process of money creation that a social contract or governmental decree does not generate money, but rather the needs of people in the economic order of the free market do. The individual market generates it by making the exchange process more fluid and by increasing the scale of transactions.

Mises has said that "money is a medium of exchange, it is the best-selling goods, and people want to use it in their future interpersonal dealings, so they want to acquire it. Money is considered by everyone to be a medium of exchange. At the same time, this is its only function. Everything else is a special aspect of this basic function (medium of exchange)".

Keynesianism and inflation will then be introduced. Inflation can be understood as "the phenomenon whereby the breaking or distortion of the rules of money production in the free market causes the quantity of money to expand beyond what it would have produced in a free market". The root of this phenomenon is that paper money, backed by the credit of the state, has been at the heart of the modern monetary system over the course of history. Paper money relies on the credit and privilege of government legislation rather than the order that naturally arises in a free market. This is the basis not only of Keynesianism, but also of the Austrian economists' attack on traditional economic theory. While in mainstream economics inflation is usually understood as a decline in the purchasing power of money or an increase in the price level, Austrian economists believe that the essence of modern inflation is to increase the quantity of money by decreasing the quality of money. In other words, people's private property is plundered by the government through the creation of new money. And the rise in prices and redistribution of wealth is a consequence of this process, not a cause.

Finally, the limitations of traditional monetary theory will be discussed. The critique of traditional economic theory by Austrian economists is mainly discussed here. We have previously described how in traditional neoclassical economics the process of production is considered to be obtained by combining a series of factors of production (labour, capital and other factors) in a way that can be calculated by means of mathematical equations. Austrian economists argue that such a quantitative relationship cannot be simply transformed by the process of production. Nor can different capital goods be completely substituted for each other. The basic fact of the real economic world is that capital is heterogeneous. The output of consumer goods can only be facilitated by a series of interlinked and complementary factors of production acting in a particular way in different stages of production. And the monetary signals of the market and the value of profits are relied upon by entrepreneurs. The optimal allocation of the different factors of production is achieved through innovations in technology and business models. In other words, a series of quantitative indicators such as aggregate demand, aggregate supply or the overall level of output of the economy does not exist in the real economic world. Austrian scholars see this as a statistical artefact created by mainstream economists, which has no relevance to real-world economic conditions. At the same time, this is the most fundamental reason for the monetary policy of inflation due to Keynesianism.

This is our discussion of traditional monetary theory and macroeconomic theory. After discussing the origins and evolution of money, we have learned that money is not only a product of indirect economic transactions, but also a product of evolution under a free economic order. Austrian economists argue that legal tender is created on the basis of government and state credit, and that it has no commodity value. Keynesianism's misconceptions about monetary theory and macroeconomics lead to inflation, which allows for blind credit expansion and policy-based taxation. This is not only an important logic of reference for our understanding of the underlying macroeconomic situation, but also an important reflection for our future discussion of monetary theory in the digital economy.

Monetary Theory of Austrian Economics

After the monetary theory of traditional economics is discussed, the monetary theory of Austrian economics will be briefly introduced from three perspectives, including the function of the monetary system, the order of the monetary economy and the monetary theory of Austrian economics. The understanding of the monetary theory of Austrian economics will not only be useful for the understanding of macroeconomics in practice, but will also be of great value in understanding the evolution of the economic order in the theory of digital economics and the core of monetary theory.

The Austrian scholars' understanding of the nature of money was discussed first, particularly with regard to the monetary mechanism. If the role of the monetary mechanism needs to be understood, then it relates to the ideas in the theory of innovation of the entrepreneur that we have discussed. In other words, the free exercise of entrepreneurial talent is seen as the most fundamental driver of market processes and economic growth. However, certain institutional conditions are required for entrepreneurs to develop their talents. Here the monetary system is the most important. According to Austrian economists, the most important institution for coordinating between different players in the market is money and the price and profit mechanisms based on it. On the one hand, a bottom-up order that satisfies a complex economy is provided by money, which enables economic calculation. If the marginal effects of economic goods in different uses are to be compared, then money is to be used as a medium of exchange. If, on the other hand, the difference in value between current goods and future goods is to be understood, it can only be understood through money, which is the equivalent of the phenomenon of interest. The monetary system can be understood as an information coordination mechanism that enables the plans and actions of the different participants in the free market to be linked in a uniform order. This process of constant reception of signals, the adjustment of one's own active behaviour and the issuing of signals is the behaviour of the entrepreneur.

The concept of monetary order in Austrian economics will be further understood. A network of people, firms and other organizations involved in the production of money is the monetary order. And above, Austrian scholars are referred to their belief that macroeconomists sum up the output of different individuals, employment and specific prices and wages in order to arrive at a set of data conclusions and systems. However, the real link between supply and demand is not reflected in this at all. Austrian economists argue that if an economy is to achieve an overall equilibrium of production and a sustainable level of employment, then different types of heterogeneous goods and services should be traded in different markets in order to balance and harmonize demand and supply, and ultimately to constitute a network based on a liberal economic order.

Here we recommend the theories of the renowned economist Rothbard. He uses the analytical framework of micro-dynamic processes to explain the process by which the monetary policy adopted by the Federal Reserve in the 1920s led to an imbalance between savings and investment and contributed to the creation of the great economic collapse of 1929 in the United States. Austrian scholars argue that the government's use of a model of intervention in the economy to influence markets caused and delayed the slowing down of the recovery process, ultimately making the consequences of the economic crisis last longer. In short, Austrian economists believe that the excessive trust in government power to secure and maintain economic prosperity and stability is a terrible thing that has disrupted the order of the free economy and amplified the impact of economic crises. At the same time, it has caused cyclical economic fluctuations.

Finally, the monetary theory of Austrian economics as well as the theory of the business cycle will be summarized. The findings of Mises and Rothbard are mainly presented here.

The role of money as a medium of credit is discussed in Mises' "The Theory of Money and Credit" [32]. Mises argues that money is more accurately understood from this perspective. At the same time, he argues that the expansion of money causes the interest rate on money to fall below the natural theory. The result of this is that when people demand money, the interest rate on loans is lower than the natural rate of interest when real capital goods are lent out, ultimately leading to an expansion of credit. According to the Bohm-Bawerk theory of capital and interest to which we have referred, producers in this case enter into a more "roundabout" production process, which makes the "roundabout" process longer and more costly for the entrepreneur. At the same time, the market receives the wrong signals, the equilibrium of the loan market is broken, fewer consumer goods are available, the market price of consumer goods rises, and eventually the phenomenon of inflation emerges.

In short, according to Mises, an economic crisis is "the loss of some of the capital invested in an excessively prolonged and circuitous production process". The disruption of the free market order by a faulty monetary policy leads to such losses and ultimately to the continuation of the business cycle. This theory was also later discussed again by Hayek in his 1931 publication "Prices and Production". Mises' theory was supported by most of the Austrian economists. They believed that the cause of the emergence of business cycles was the excessive supply of money, the essence of which was that the expansion of money led to a systematically lower money rate than the natural rate.

Rothbard's "Man, Economy and State" [33] covers almost all of his findings on monetary theory. The book's theories on exchange and prices will be. Unlike the classical economists who usually discussed production before exchange, Rothbard discusses exchange first and then production. The underlying reason for this is that Rothbard saw exchange as the basic logic of a free market economy, and the basic premise of exchange as being associated with 'justice'. At the same time, Rothbard saw economics as a science of 'exchange' behaviour, rather than a study of 'material goods'. This view is in line with our discussion of the idea that Austrian economics is a science that studies the actions of people to satisfy their desires. Different people place different subjective values on different goods, and this makes the act of 'exchange' appear. However, when there are no goods to exchange, the act of production is triggered, and prices are generated on the basis of exchange.

The concept of the value scale was used by Rothbard as an alternative to the supply and demand curve to describe the nature of prices. He believed that prices lie between the highest purchase price of the most capable buyer and the highest price of the second most capable competitor. In other words, the price of a property is determined by the marginal buyer and the marginal seller. Rothbard suggests that the stock of money and property and the retention of money and property are the ultimate determinants of the price of demand. When the demand for inventories of possessions and retention of money increases, the price also decreases.

This theory gives us an insight into the process by which money is created and how its true value is manifested. In particular, there is no objective value for property, as advocated by Austrian economics. There is only a price obtained by different people playing in a complex market, and the price itself contains the value judgement of each individual. Therefore, the role of money is reflected in the expansion of specialization and the division of labour, which allows the limits of specialization to be explored by individuals. Based on this theory, Rothbard developed the "Angel Gabriel" model to illustrate the theory that the quantity of money is not important. This model is loosely interpreted as the intersection of the demand curve for the total stock of money and the cash balance of the society, which determines the price of money (purchasing power). At any given moment, every unit of money is owned by an individual, so the total supply of money will always be exactly equal to the sum of individual cash balances at that moment. In short, the market always tends to operate in such a way that the ultimate situation of the purchasing power of money is determined by a consistent equilibrium between the supply and demand for money. The amount of money available within the economy is always sufficient to enable everyone to get everything with it.

In this chapter, the monetary theory of Austrian economics is discussed in terms of the origins of money, its institutions and its economic nature, as well as reflections that differ from the monetary theory of mainstream economics are given. The different roles of money in the natural order and money as a medium of credit are understood. At the same time, it is the reality of resources, labour, capital and commodities that really represent the wealth of society that is understood. Thus, when money increases, society does not become richer, but rather it drives economic cycles, as well as driving up the prices of different goods and services. In other words, a given economy does not determine the amount of money in the market, but rather its should be determined by different entrepreneurs of money in a dynamic market process. We need to understand here not only the logic and rigorous thinking of monetary theory in Austrian economic theory, but also the revelation of the nature of markets. This has value for future discussions of the monetary theory of the digital economy. At the same time, this theory has limitations in a real market. There is no doubt that the credit of governments is valuable for money. At present, the market being purely dependent on driving the money supply system is not mainstream and does not correspond to reality.

Niche Economies and Token Economy

The creation of the virtual currency Bitcoin was a landmark event for blockchain technology, and the financial crisis of 2008 was the backdrop for Bitcoin's birth. This is something that is well known. Dorian S. Nakamoto, whose is the founder of

Bitcoin, left a quote on the foundation block of Bitcoin, whose was the Chancellor of the Exchequer on the verge of implementing a second round of bank emergency bailouts on January 3, 2009. The quote was also the front page headline in "The Times" that day. This first set of bitcoins, whose number was 50, was officially born. Bitcoin has an important historical context and significance, referring to the hope that the problem of periodic crises rooted in the modern financial system will be solved by the creation of a cash system built on peer-to-peer technology that is not controlled by central banks or any centralized financial institutions. How then should the problems of the modern financial system, and the key role played by money in it, be viewed? It is important to look at the development of the international monetary system to understand the background to the birth of digital money and the current state of money in modern finance capitalism.

US President Richard Nixon announced to the world on 15 August 1971 the end of the exchange between the US dollar and gold, its a landmark event for the global monetary system. At the same time, he announced the collapse of the Bretton Woods system. From this event it can be concluded that the monetary system of mankind and the system of material commodities completely disintegrated. It was also the first time that mankind entered the age of full credit money, also known as the age of fiat money. Until then, a material commodity with a stable value had always been used as a reference for fiat money values, such as the gold standard, the silver standard or the bimetallic standard. The era of credit money. Sovereign governments were given the power and responsibility to manipulate money by credit money, which led to national deficits, the issuance of national bonds and the possibility of indirect taxation through inflation. As a result, a series of modern central bank monetary policies were born, such as the regulation of the money supply, the regulation of interest rates and the control of liquidity. On the other hand, the foundations of the fixed exchange rate system were destroyed and floating exchange rates became the norm in national monetary relations. This made the international financial system extremely fragile and sensitive, and the 2008 financial crisis was caused by the creation of fiat credit currencies, the divergence between the virtual and real economies in the global financial system, and the over-inflation of global capital in the modern economic development process. Eventually, the birth of Bitcoin and the mechanism of digital currencies with the token mechanism at their core was driven.

Digital asset exchange Gemini Trust and blockchain company Paxos Trust were approved by NYDFS on 10 September 2018 to each issue a stablecoins pegged to the US dollar (GUSD and PSD). This has received a lot of attention from academics and industry players in the digital economy, particularly in relation to blockchain financial innovation. The exchange between stablecoins and other cryptocurrencies is very convenient and challenges to the regulation of traditional cross-border capital flows have developed. As a result, there is greater industry interest in it. At the same time, an outlook on the financial and business changes brought about by blockchain technology has been derived. In this chapter, the intrinsic logic and context of digital currencies will be explored in depth based on the perspective of digital economics. At the same time, the formation of a new economic order based on the concept of "niche economies" is also presented to provide a new perspective on the understanding of blockchain technology and digital currencies.

Here are the conclusions. There is no purely technological revolution that can provide a decentralized monetary solution at a level beyond the state for the foreseeable future. At the same time, the existing monetary policy and economic order cannot be challenged for no reason whatsoever. If the monetary and financial order in the macroeconomic sense is to be changed, a supranationalist political and economic solution based on social consensus needs to be established. Innovators cannot drive this solution from the bottom up. In the commercial marketplace, a new value incremental network based on the transformation of business organizations from 'contractual organizations' to 'niche economies' and delivered within the framework of the existing economic order is the real business path that will enable the development of blockchain technology to be driven. In other words, the change in overall economic value and order is driven from the bottom up through the business transformation of enterprise organizations to enable the digital transformation of economies and the implementation of large-scale commercialization of blockchain. This falls within my understanding of the inherent meaning of the revolution in digital economic order brought about by blockchain technology. This topic will be divided into two parts, which include the inherent paradox of stablecoins, niche economies and the new digital economy order, and token value and cryptocurrency theory.

The paradoxes inherent in stable currencies are discussed first. When blockchain technology is discussed, the consensus mechanism is the most discussed. That is, the same tasks and resources are shared through distributed computers in order to allow communication and computational problems between networks to be solved. A consensus mechanism refers to a technical mechanism applied by blockchain or distributed ledger technology that does not rely on a central authority to authenticate and validate transactions. It is the foundation of all blockchain and distributed ledger applications. Below are three important phases in the development of consensus mechanisms.

In the first phase, two computer scientists, Leslie Lamport and Barbara Liskov, winners of the great Turing Award, developed the classical consensus protocol BFT in 1982, with its "permission voting, majority rule" approach to consensus building. Features. It also has the advantage of fast settlement and fast secured transactions. In addition, it has poor scalability and can only be used in a limited number of network nodes. This is one of its disadvantages.

In the second phase, Dorian S. Nakamoto proposed the POW protocol in 2009, which is a representative event. Extremely scalable networks were formed without all nodes needing to be known. This is its strength. However, this comes at the price of extremely low computational efficiency and very limited throughput.

In the last phase, POS and its derivative mechanism represented by POS provided by Ethereum and PBFT provided by Hyperledger are proposed. At its core, the need for high speed transactions and other scalability requirements are realized through more flexible and complex technical mechanisms (e.g. voting mechanisms, etc.).

After these three phases, the technical basis for the application of blockchain commerce is taking shape and is gradually being implemented in financial and non-financial markets.

Next, the inherent logical paradox of consensus protocols and digital currencies will be discussed. In fact, the original idea behind the creation of stablecoins was that the problem of excessive price volatility in cryptocurrencies would be solved by means of "collateralised credit". Here are the three main models of current solutions.

The first is the model represented by USDT, which means that the exchange of legal tender and USDT is achieved by depositing the same amount of US dollars in an official account.

The second refers to the issuance of stable currencies through digital assets instead of US dollars as collateral, represented by MakerDao.

The third refers to a market where the relationship between the token process is regulated algorithmically so that the price of the token is stabilized at a fixed ratio to the legal tender.

In fact, either model requires the establishment of indirect credit to achieve a "stable currency". According to the monetary theory of digital economics, all currencies are not based on "algorithmic credit", but on "commodity credit based on access". Thus, an inherent paradox of the existing digital currency ecology is argued by the emergence of stablecoins, which refer to the creation of decentralized monetary mechanisms that allow the credit expansion problem of the existing macro monetary system to be solved. At the same time, credit links are attempted to be realized through the establishment of credit links with the existing fiat money system. On the one hand, it is hoped that technologies such as consensus algorithms will enable the decentralization of money and the overall limitation of the number of issues, so that the problems of monetary stability in the digital economy can be solved and the consequences of excessive credit expansion in the real economy can be avoided. On the other hand, an artificially intervening monetary policy is designed in such an ecology, and the problem of "commodity credit" for digital currencies is solved by the creation of indirect credit. There is no doubt that such an approach does not have a beneficial effect. Not only does it not allow the credit problem of digital currencies to be solved, but it also does not allow the problem of currency stability to be solved.

When technology in the digital economy is understood, the changes it brings are often overestimated, and the institutional elements behind the technology are ignored. Indeed, the institutional study of monetary policy in the digital economy is currently under-appreciated. The creditworthiness of any currency is also dependent on the balance of the elements of its economic ecology rather than on changes in a single element. Therefore, the existing stable currency mechanism is not the ideal monetary solution for the digital economy. Since the inception of money, attempts have been made to seek relative stability in the value of the currency, and this has never been interrupted. To date, the gains have also been relatively limited. There is no market for "stablecoins" without a reconfiguration of the institutional elements.

Based on the above analysis and judgement, the following questions need to be pondered. What is the real value of the transformation of the relations of production in the digital economy brought about by blockchain? Therefore, the topic of niche economies and the new order of the digital economy will be discussed. First, I will present its conclusions. I believe that for the future of the digital economy transformed by the blockchain, the expansion of businesses from simple production and market sectors into an ecology of niche economies will be realized through the issuance of digital token within business organizations, enabling the scaling up of business value and the reduction of transaction costs across the market.

Here are three layers of what a niche economy is.

The first layer refers to the transformation of the business organization from a simple market player in production transactions to an economic ecology capable of an endogenous growth logic. In other words, it becomes a dual structured organization with not only corporate but also market attributes in the digital economy market.

The second layer, the core of the business organization is not only production, but the creation of an economic ecology for the entire eco-community. On the one hand, if the differentiated needs of each individual are to be met through the production of goods and transactions, then a "community" economy in the broadest sense needs to be established. On the other hand, the value proposition of the user in terms of "customized and differentiated needs" has to be met through the enhancement of the entire ecological experience.

The last layer, niche refers not only to the size of the users, but more precisely to the users based on different value pursuits. An "imaginary community" is to be constructed by the organization in order to enable products and services to create higher value for a specific group of people.

The fact that only the "niche economies" within the digital economy make sense is noteworthy. The "experience economy" and the "service economy" are provided by the digital economy ecosystem. Thus, three different conclusions can be deduced in order to make sense of the changes in business logic brought about by the blockchain.

According to digital economics, the digital economy is a market mechanism centred on the right to use, rather than ownership. Therefore, the trading of usage rights is the essence of property rights trading. And niche economies have to create differentiated experiences and services to suit each individual experience. An efficient transaction network mechanism is provided by the blockchain to enable the entire credit ecology of right-to-use transactions to be secured.

According to digital economics, niche economies have dual properties, which include the properties of a firm and a market. Therefore, the unit of measurement of value created in the internal process of value exchange is the essence of the token issued by the niche economy. Therefore, the most important role is how the token enables the internal market to function more efficiently and stably.

According to digital economics, the existence of niche economies necessitates the alignment of industry analysis and enterprise analysis. Firms need to not only offer specific product services but also consider the creation of ecological capabilities in niche industry sectors. When there is a fuller division of specialization, the greater the potential for expansion of this industry. This is what Adam Smith has taught us. Therefore, the fundamental question that niche economies should address is how to create some new industries to provide better services to users in their own ecology.

Once the above three basic logics are understood, the implementation scenario of blockchain technology can be discussed in the context of the "incremental" real economy, instead of being limited to the plundering method of the real economy's advantages in the stock, or the "virtual economy" represented by finance. The "virtual economy", represented by finance, has changed its basic logic. On the one hand, the digital economy is completely different from the ecology of the real economy, whose goal is the creation of a "re-tribalized" niche economy. On the other hand, most digital economies do not yet have sufficient commodity credit. Therefore, the problem to be solved at this stage is how to accumulate value and build consensus within the niche economies, rather than blindly challenging the existing financial order.

Finally, the relationship between the value of a token and a currency is discussed. In fact, the term token predates the blockchain, and the token economy was applied by Yale professor Alan E. Kazdin in the 1880s in the fields of psychiatry, clinical psychology and education. He also published a monograph, The Token Economy: A Review and Evaluation, in 1977. He is not only the 'Stirling' Professor Emeritus of Psychology and Child Psychiatry at Yale University, but is also the Director of the Yale Parenting Centre and Child Behaviour. In addition, he served as President of the American Psychological Association in 2008. When the concept was introduced, it did not cover the economic component, but rather the 'incentive' component was more discussed, which refers to the behaviour of others being guided by incentive behaviour. Currently, the discussion of token is more on the economic side. The relationship between token and monetary theory will be analyzed from the following two perspectives.

From the perspective of circulation of value, a niche economy pass-through can be considered as a digital credential of internally circulating crypto-digital rights. It does not have the role of a legal tender currency in the narrow sense, but only the role of circulation within the ecology of the network organization. In other words, it has the value of an internal ecological currency based on the "right to use". According to Mises' "retrospectivity theorem", the value of money is ultimately retrospective to its price as a commodity of exchange of material. Thus, a token based on the exchange theory of "right of use" has a "monetary value" in some sense. In this theoretical framework, a good can be understood as a medium of exchange in general on the basis that it is itself derived from people's subjective evaluation of its utility. At the same time, it can be understood that the discussion of the monetary value of the token economy is constructed on the basis of a theory of access.

Derived from the perspective of incentives, tokens can be understood as a form of incentive in a free and open collaboration mechanism created in crypto economies. The creation of a publicly visible distributed ledger is the essence of blockchain technology, which is used to record the historical information of transactions. In the process, it enables the 'mining mechanism' of tokens to be used to incentivise and punish corresponding behaviour. This is the key to the understanding of its incentive mechanism. Through the consensus-based, scalable, Turing-complete applications that blockchain technology can create, anyone can further reduce the transaction costs associated with contractual incompleteness within niche economies through freely defined ownership rules and transaction methods set out in smart contracts, ultimately enabling the value of tokens to be realized.

In summary, the monetary theory of digital economics is discussed through three chapters, which include the monetary theory of traditional economics, the monetary theory of Austrian economics, and niche economies and When the nature of money and the concept of niche economies are understood, the value of tokens and the internal logic of blockchain commercialization will not be blindly understood. We are not only must not confuse tokens with the logic of digital currencies issued by central banks, but we must also not lose sight of the true value of tokens. This is the knowledge we should have for the study of the monetary theory of digital economics. Monetary logic and monetary theory are important topics in the field of macroeconomics originally. The reason why it is now placed in the field of digital economics is that it is analyzed on the basis of the logic of companies and industries, rather than being studied from the perspective of the substitution of legal tender values.

Chapter 12
Institutional Economics in the Digital Era

In this chapter, institutional economic theory in the digital age will be discussed, with the main focus on the evolution of institutional economic theory in the digital economy. Here are three basic reasons why this topic is being discussed.

Firstly, the study of economic behaviour is based on the changes in human social institutions. Therefore, the relationship between social institutions and economic development needs to be understood first and foremost.

Secondly, the contribution to the change of business systems is one of the important contributions of blockchain technology, where the economic principles refer to the understanding of Coase Theorem. This part needs to be discussed in more depth.

Finally, self-organized communities across geographies and countries are an important feature of the digital economy era. Institutional economies and governance mechanisms in this dimension need to be discussed.

For these reasons, the discussion of institutional economics in the digital age is divided into three parts, which include economic and social institutional changes, smart contracts and Coase Theorem, and institutional change and the digital economy.

Economic and Social Institutional Changes

If the relationship between economic and social institutional change is to be discussed, it needs to be framed in terms of the historical evolution of human societies, and D.C. North, winner of the Nobel Prize in Economics, co-authored "Violence and Social Order" [34] with John Joseph Wallis. In this book, the growth of national economies is examined in depth. They found that the long-term development of national economies cannot be explained by capital accumulation, resource endowments or geographical factors. However, national economic growth can be explained by the institutional arrangements that are formed in different societies,

Z. Liu, *Principles of Digital Economics*, Contributions to Economics,
https://doi.org/10.1007/978-981-16-9020-4_12

especially economic institutional arrangements. Its underlying logic is that people regulate human behaviour through institutions, and the complex social interactions that take place through organizations. Thus, if the process of human development is to be understood, then how institutions shape the organizations that enable sustained human cooperation needs to be understood. The relationship between economic development and institutional change will be discussed through D.C. North's study of social development as well as institutional change.

(1) North argued that the concept of social order is formed when the economic and social systems are placed within a framework. The primitive social order refers to pre-written societies. (2) Limited access order refers to an order in which rents are created through restrictions on access and violence is regulated by a political system at its core. It refers mainly to the order of the feudal state before modern times. (3) Open access order refers to the order created through political and economic competition. It mainly represents the order of the modern state. Among the open access order, institutionalized competition, free entry and mobility and the long-term economic development resulting from the free order of the market are the most important features. North argues that there is a tendency for both political and economic systems to become increasingly open. At the same time, he argues for the logic of this through the theory of double equilibrium. That is, both economic and political systems either tend towards openness or both tend towards restrictions of access. If there is no fundamental change in one system, then there will be no fundamental change in the other. So how do you explore an open economic system by constructing an open political system? This is what North sees as an important question for the study of social development and the transformation of systems. And this part is certainly very important for the study of institutional design in the digital economy.

It is worth noting that the internal logic of institutional change and economic growth in North's "Institutional Change and Economic Growth" is analyzed. The theory of institutional change allows a group of actors (individuals or groups) to see the benefits that a new institutional arrangement will bring and to compensate for the costs of innovation in the form of organization, particularly in terms of potential economies of scale, reduced information costs, risk diversification and the internalization of externalities. This is how he sees the relationship between institutional change and the economy. Furthermore, there are two basic forms of institutional change. One relies on external coordination by government and the other relies on the forces of spontaneous order for institutional innovation. And the consequences of different institutional innovations can be divided into two types, which include productive institutional arrangements and redistributive institutional arrangements. That is to say, institutional arrangements that may lead to economic growth or to redistribution of stocks. This is the subject we need to focus on when designing the relevant institutional arrangements for the economy.

Next, North's study of the evolution of human social institutions and limited access order [35] will be discussed. Priority access order has been the most dominant institutional arrangement in the development of human society in the past. Institutionalized rent setting, market power, privilege, and the distinction between elites and citizens are its characteristics. At the same time, North sees the relationship between politics and economics as one in which government acquires economic rents through restrictions on economic access so that credible commitments are entered into among the elite, ultimately enabling existing securities to be backed and social order to be provided. The political system is not independent of the economic system. The main player in economic activity is the political system, and political relations are built on economic rents. Thus, North argues that without open access to organizations, then neither a competitive economic society nor a political society would exist. The key to the development of modern society is the realization of open access to social organization. In short, the process of economic development is a transition from the Priority access order to the open access order, and the important conclusions of the new institutionalism include this conclusion. The disciplinary indexing tools of the new institutionalism (such as game theory and statistics) are combined in the study of economics, politics and society as a set of analytical tools. This set of tools has provided important insights into the understanding of the relationship between institutions and the economy.

According to North's theory, the effects brought about by institutional arrangements will be briefly analyzed. North mentions the following two fundamental questions about institutional analysis. Firstly, how is the basic institutional environment shaped and changed? Second, what are the consequences of any set of basic rules of decision-making? North concludes, through the process of studying constitutional change in American history, that its a change in the Supreme Court's interpretation of the Constitution that usually leads to relatively lagging economic benefits. The basic logic here is as follows.

First, when the gains from institutional change increase, the stimulated institutional arrangements and innovations will be induced and stimulated at a lower cost than changing the underlying institutional arrangements. It will generate attempts to circumvent the limits of the principles of the underlying decisions to make profits. In other words, the institutional arrangement cannot be perfect and the secondary institutional arrangements it brings about will cause transaction costs to be raised and even violations of the principles of the underlying decisions to be induced. Thus, changes in institutional economics cannot only bring benefits, but also costs.

Secondly, when the principles of fundamental decision-making change, so does the redistribution of wealth and income and changes in the patterns of fundamental political power. So future predictions are just not accurate. And the outcomes of institutional change can be judged. And outcomes outside of economic growth must possess the risk of uncertainty.

Thirdly, its impact on economic growth is determined by the qualities of the institutional environment and the extent to which the new system is introduced. If the regime places more emphasis on increasing output and takes into account the new rules of income distribution, then economic growth can be expected to be achieved. Conversely, if only income distribution is discussed and economic growth is not achieved, then the institutional arrangement will end in failure.

The implications of this chapter for the study of digital economics will be discussed at the end. From the two topics discussed above, on the one hand, from the perspective of social development and institutional change, North once said that "social science studies how human beings interact with each other to produce the complex social structures in which we all live. The focus of research must therefore be on organization. We need to understand how different groups of people organize their relationships in long-term, and ultimately perpetual, interaction". Therefore, the understanding of organizations is crucial and important in digital economic. And institutions are key to understanding how organizations are formed and how they act.

Smart Contracts and Coase Theorem

The use of smart contracts is one of the most important technological innovations in the development of blockchain technology. The concept was developed by Nick Szabo in the 1990s. It is almost the same age as the Internet (in fact, the concept of the Internet was not popularised until after 1996.) Ethereum first saw the fit between blockchain technology and smart contracts in 2013. At the same time, a white paper entitled "Etherium: The Next Generation of Smart Contracts and Decentralized Application Platforms" was published and the concept of smart contracts was resurrected. So far, it has grown to become the top institution in the smart contracts space. However, the well-known technical aspects of smart contracts are not discussed here, while the economic implications of smart contracts and their value for enterprise theory will be discussed. Blockchain is being treated as the new revolution of the Internet. The impact of this revolution needs to be considered not only in terms of the technical aspects of the innovation, but also in terms of the impact of the ideas arising from this technological innovation on the future economy. Issues related to property rights have been discussed before, and here it will be analyzed again in terms of institutional economics and organizational change.

Specifically, I believe that the concept of "smart contracts" is in fact a new interpretation of the organization of industry and enterprise in the network economy. In the future, firms in the network economy will become increasingly marginalized, while the consensus communities built by the network will become the mainstream. We need to interpret this theory in the context of the discussion of firms and markets in economics, particularly the chapter on the theory of the firm. First, the current dimension of economists' understanding of markets is understood through a discussion of Ronald H. Coase's and Zhang Wuchang's views on firms and markets. We then discuss why the functions of firms in traditional markets will be replaced by network organizations formed under the mechanism of smart contracts.

The Coase Theorem is not only the basis of the economic theory of property rights, but also one of the basic logics for the analysis of economic problems. The relationship between this law and the blockchain economy is first explored. This law can be understood from three perspectives. Firstly, when the cost of a transaction is zero, the initial definition of rights does not matter. Secondly, in the case of transaction costs, the definition of property rights facilitates efficiency. Finally, the initial power is precisely defined through the government, which will be preferable to a transaction between private parties to have the initial allocation of rights corrected. In short, transaction costs can be reduced by the definition of property rights, so that market efficiency is increased. The essence of this is the discovery of the relationship between transaction costs and property rights arrangements, which refers to the impact of transaction costs on institutional arrangements. This theory is used to analyse the consensus mechanism in the blockchain economy, which is essentially an institutional arrangement. Based on Coase Theorem, any kind of institutional arrangement has a corresponding cost. At the same time, all institutional arrangements are affected by the creation and change of transaction costs. Therefore, the essence of different consensus mechanisms (POW, POS or others) being adopted in blockchain is that transaction costs are reduced through institutional arrangements so that the value network can be operated more efficiently. This is an important perspective from which future blockchain communities or DAOs (distributed self-organizations) will be understood. The future economic form of blockchain will replace the enterprise with a network so that higher transaction efficiency and lower transaction costs are obtained.

Therefore, I believe that Coase Theorem is worthy of being debated on the following theory. Coase argues that firms (corporations) are the only way to reduce transaction costs. And this is limited by the information revolution, which occurred when Coase was writing on his thesis. It was also influenced by the fact that the network economy was not yet developed. In "The Nature of the Firm" [31], Coase argues that the role of the market has been replaced by the firm. The nature of the firm is a substitute for the price mechanism, which costs money to operate, and the market also incurs costs to operate. The firm replaces market coordination through management coordination. At the same time, the enterprise is the inevitable result of cost reduction. Therefore, firms exist. In addition, he argues that markets and enterprises are interchangeable means of allocating resources. The difference is that the allocation of resources in the market is regulated by the price mechanism, while within the enterprise it is regulated through the management mechanism. From the economist Zhang Wuchang's analysis of the contract economy in "The Economy Explanation", it is clear that the market is considered to be sufficient only for contracts, and that the firm is a special form of contract. He suggests that "the firm is a form of contractual arrangement" and that the substitution of the market by the firm is simply "one type of contract for another". At the same time, he argues that consensus is not defined in Coase Theorem, and that product markets and market elements are not clearly defined.

More precisely, there is no clear definition of what is meant by a company. On a smaller scale, a company can be a contractual relationship between two asset owners. On a larger scale, a company can include an entire economy. The allocation mechanism of the market is understood by Zhang Wuchang through the concept of "contract". He believes that all individuals are linked by contracts. Traditionally, income distribution and resource allocation have been the focus of economists, while contractual arrangements have been neglected. In a complex economic world, the subject of a complex real economy is in fact transaction costs and the conclusion of contracts. There is a corollary to Coase's law about zero transaction costs in perfectly competitive markets, and it is even more unlikely to be realized. I therefore think that Zhang Wuchang's idea should be adopted. He argues that firms are only one type of contract. At the same time, he argues that it is the result of a long-term contract called a "firm" being used as a substitute for a series of short-term contracts in a "market" to save transaction costs.

Based on these conclusions, a new inference is made as to whether, if a contractual form exists that is different from the market and the enterprise, it can replace the role of the enterprise and the market in resource allocation, or whether it can fulfil both roles. Clearly, the answer is that in a smart contract-based blockchain economy (network economy), a "new contract" can be used to replace the "special contract" of the firm, so that the function of resource allocation is fulfilled. In other words, in traditional market and firm theory, the only answer to the problem of the costs of market transactions is the firm. In network economics, however, a better market mechanism than the firm form is the smart contract.

Logically, the problem of the rational form of contracts is actually solved by smart contracts in the blockchain economy from three perspectives. Firstly, the logic of flatness and almost zero transaction costs can be realized through peer-to-peer transactions, which are a natural advantage of a decentralized economic form. Secondly, the establishment of trust in market contracts can be automated by smart contracts for consensus mechanisms. The efficiency of the technological contract exceeds the efficiency of transactions guaranteed by the legal system. Finally, the transmission of network value is achieved through the mechanisms of communities. The efficiency of transmission in a blockchain community far exceeds that of the traditional corporate hierarchy. Thus, Coase Theorem fails in network economics or blockchain economics. In the future of organization, the enterprise is no longer a given.

Finally, contract theory and the network economy approach are used to discuss the value of digital economics. The value theory of digital economics that we have discussed not only gains new theoretical explanations under the logic of property rights theory, but also leads us to think about the field in a new way.

First, the sharing economy is the essence of the network economy. As such, the scope of property rights definition shifts from being possession-centric to being usage-centric. This is where I believe the biggest transformation of the blockchain to the physical industry will exist in the logic of the sharing economy. In classical business discussions, asset exclusivity is very common. At the same time, new costs arise due to the issue of "squeezable quasi-rents" embedded in the assets. "The term 'exclusive quasi-rent' refers to the part of the contract that is extorted after the conclusion of the contract because of the existence of asset exclusivity". However, it is predominantly a right of use in the theory of property rights in the digital economy. Thus, the problem of asset exclusivity does not exist and smart contracts contribute to the solution of the "exclusive quasi-rent" problem.

Secondly, the network organization (e.g. DAO) is the main body of the blockchain economy. This structure possesses not only the managerial properties of a business, but also the transactional properties of a market. The value of the network must therefore be discussed in terms of both the managerial and market efficiency dimensions of the ecology. At this point, theoretical homogeneity is obtained by management and economics. Based on the system of property rights, the exchange and cooperation between people refers to the contracts concluded by property rights. In other words, the main elements of the market economy are constituted by contracts. A new type of contractual arrangement is provided by networks, which combine the properties of firms and markets.

Finally, the discussion of rents under generalized equilibrium theory and generalized Pareto is necessary for the blockchain economy. In other words, the social capital of the network economy derives its profits not only from the filling of structural holes by entrepreneurs, but also from the rents generated in the process of reallocation of market resources by innovation and individualization, and not from differences in corporate resources or the oppression and exploitation of a particular group. It is when these three basic economic logics are understood that the subjects in the new economic ecology can be understood. In other words, the theory of a blockchain economy based on network economics can only be derived when the question of how a network based on smart contracts generates value is understood.

Institutional Change and the Digital Economy

After the arrangements for institutional governance at the firm level and the principles of economics have been discussed, the relationship between institutional change and the digital economy will be analyzed in depth. At the same time, the analysis of this theory will take the community governance mechanism of the blockchain as a case study.

Firstly, we need to be clear that the traditional neoclassical economic theory of modern institutional economies has usually been neglected. When economic models are constructed, clear property rights, frictionless information and virtually institutionally cost-free transactions are usually assumed. In neoclassical economics, the market is considered to be the only efficient and rational mechanism for institutional arrangements. However, states, societies and firms actually have unique institutional arrangements. And the efficiency of resource allocation and the transaction costs of synergies between different individuals are determined to a large extent by the efficiency of these institutional arrangements. The reality is that there are various institutional arrangements that differ from the market mechanism. In other words, any society is faced with different institutional choices and the possibility of institutional change. The common code of conduct to which members of society adhere is the institution. It aims to be a way for an organization or society to deal with future uncertainty and to enhance the usefulness of the organization. From this perspective it follows that organizational ecologies and services of particular attributes can be provided by any regime. And for any institutional arrangement, the inherent transaction costs need to be understood, especially the costs associated with institutional change. This is the logic underlying our study of the theory of institutional change in the digital economy, which refers to higher efficiency and lower costs being obtained by the entire organizational ecology of the firm through technological means, in order to obtain more efficient institutional arrangements.

Differences in institutional arrangements and institutional structures are then to be distinguished. At the same time, the application of these two approaches in the digital economy needs to be understood. It is the codes of conduct that bind people to particular patterns of behaviour and particular relationships that are institutional arrangements, such as families, businesses, universities and governments, to name but a few. And institutional arrangements include both formal and informal arrangements. There are various types of institutional arrangements in the digital economy, such as borderless organizations based on the sharing economy and distributed organizations based on the crypto economy, to name but a few. Furthermore, it is important to be clear in which areas different institutional arrangements are applied and that there are deep interconnections between different technological paradigms and different institutional arrangements. Institutional structure refers to the sum of formal and informal institutional arrangements in a society. For example, in the digital economy there are "central" and "decentralized" institutional structures and "traditional" and "intelligent" institutional structures. When we talk about institutional change, we are referring to changes in institutional arrangements rather than changes in institutional structures. Institutional arrangements are ways of making the whole organizational ecology more profitable. Institutional structure, on the other hand, is the way in which the evolution brought about by the social ecology and technological paradigm results in structural changes in the entire economic and social machinery.

In the digital economy, the holistic changes brought about by technological paradigm shifts can be understood as an inherent evolution of institutional structures. This evolutionary process is relatively slow. So far, there have only been two relatively obvious structural changes as follows. The first was the change in institutional structure brought about by the rise of the internet economy. The industrial organization ecology and management ecology represented by the Silicon Valley companies was recognized by all internet entrepreneurs worldwide. At the same time, flat, innovation-centred and product experience-oriented internet organizational arrangements became the typical phenomenon of institutional structural change. The second time is what we are experiencing. The algorithmic economy, represented by the sharing economy and the blockchain economy, has brought about the underlying logic of the institutional structure of the entire organization with efficiency and security at its core, which is constituted by a decentralized and distributed organization. So far, the transformation of this institutional structure has not been completed. At the same time, it is going through a phase of expansion of the scope of the sharing economy and the gradual exploration of the blockchain economy. There is no doubt that the underlying logic of the institutional structure or institutional arrangement is therefore security and economy. Both of these elements need to be balanced in any institutional arrangement. In the case of the sharing economy model, if only the expansion of economic scale is taken into account and the bottom line of security is not, then something like DIDI will pay the price and risk in terms of some security issues. Ultimately, the very survival of the business will be threatened. Therefore, the synergy between security and economics cannot be ignored when the institutional structural changes caused by digital technology are taken into account.

Finally, the application of institutional change theory in the digital economy is analyzed using the community governance mechanism of blockchain as a case study. This is because blockchain technology not only represents a transformation of people's collaborative relationships in the digital economy, but also drives the transformation of the entire corporate ecology. Just as the industrial revolution drove the transformation from an agrarian to an industrial capitalist society, blockchain technology is not only a technology for organizational change, but also offers the possibility of a more decentralized social contract. The issue is therefore discussed from three perspectives, which include distributed ledgers and social institutions, the dual governance structure of the blockchain economy and the autonomous ecology of blockchain communities.

The relationship between distributed ledgers and social institutions is discussed first, and its the focus of an understanding of the relationship between technology and institutional change in the digital economy. The concept of the ledger is not just a tool for recording accounting transactions, but a record of the contracts that are formed when people agree on something or a fact. The ledger confirms the rights, obligations and property arrangements of the parties to a contract through a record of their ownership, identity and different states of affairs, so that different economic institutional arrangements for different social collaborations are formed. Modern double-entry bookkeeping techniques can be considered to be the basis of the modern capitalist economy. Oliver Williamson, winner of the 2009 Nobel Prize in Economics, suggests that differences in transaction costs between these institutions determine the behaviour of people in production and exchange between markets, firms or governments. And the way and efficiency with which different institutions manage their books is determined by this difference. Maintaining information on the books of citizens and employees is an important task for modern governments and businesses. The distributed ledger technology offered by blockchain provides a new ledger technology for the crypto and digital economies to enable the evolution of economic ecosystems to be driven by game theory and incentive design.

Distributed ledgers need to be looked at, with what are in effect new ways of being used to record property rights relationships. For any society needs to solve the problem of conflicting interests between its social members (it is the problem of competition). The problem that economics has to solve then is how to allocate rights to individuals through the definition of property rights. Here we want to emphasise that private ownership is not negated by the fact that blockchain promotes public ownership. In fact, the difference between public and private ownership is that the owner of the commons cannot sell the shares of the commons that belong to him or her. In contrast, in a blockchain economy, when an exchange of value takes place, the transaction needs to be carried out through a pass-through. The definition of property rights remains very much a central issue in blockchain networks. Property rights are essentially a social contract to help the economic community form rational expectations. Blockchain has only changed the way in which such property relations are formed and the institutional arrangements for attributing the value of property rights. In the blockchain economy, property rights that previously needed to be achieved through social contracts can be achieved through technical contracts. In addition, the revenue arrangements that were previously obtained through the shareholding system can be realized in the blockchain economy through the pass-through economy. This is the basic logic of our understanding of the institutional economy formed by blockchain. It is still a product of market mechanisms, not of a utopian commons.

Then the dual governance structure of the blockchain economy is discussed. A decentralized, de-trusted, autonomous, information immutable mechanism is constructed by the distributed ledger provided by the blockchain. At the same time, such a community needs to be governed. Then a dual governance structure will undoubtedly need to be used to manage the community. On the one hand, the ownership of the ledger and the overall ecology of economic incentives is confirmed through algorithmic and arithmetic-based mechanisms like POW (Proof of Work). On the other hand, efficiency and community control can be enhanced through voting mechanisms such as PBFT (Byzantine Fault Tolerance Mechanism) to enable issues such as forking and performance in the blockchain ecosystem to be addressed.

In other words, a dual structure that balances efficiency, security and performance can be created through the technical mechanisms of the blockchain. Security issues are addressed by one of the institutions. The ecological security within the governance system is guaranteed through oversight to prevent the entire ecology from being evil in the process of decentralization. Another body is responsible for the efficiency and performance of the entire governance system in order to maximize the efficiency of the security of the technical ecology. This is a decentralized, dual governance structure based on a voting mechanism and proof-of-work mechanism that we provide. The advantages of blockchain technology have the potential to be maximized by such an organizational structure and institutional arrangement to enable a more secure and efficient economic ecology to be constructed.

Finally, the impact of the autonomous ecology of blockchain communities on organizations such as banks and corporations, and the regulation and control of the risks involved are discussed. Here, the previously discussed cryptoeconomics needs to be examined in conjunction. On the one hand, we need to understand that the modern capitalist system has evolved based on ledgers. Governments maintain a ledger of citizens' data and corresponding rights, businesses maintain a ledger based on assets such as employment and property rights, and banks maintain a ledger of the monetary assets associated with savings and investments in society. Thus, a completely new bookkeeping technology is offered by the institutional economics of blockchain in order to enable a new governance mechanism to be offered. On the other hand, a highly secure and trust-free institutional economy is offered by the cryptoeconomics of the blockchain. In other words, the conclusion of technological contracts is driven by the design of games and incentives in order to allow a whole set of organizational ecologies and governance structures to be developed in line with the logic of the digital economy.

While the benefits of blockchain's institutional economy are clear, there are also risks that need to be seen. For example, crypto banks fulfil the transactional aspects of money through smart contracts in order to allow the regulation and control of organizations to be completely disconnected. This is a very big shock to the existing economic order. And the immaturity of blockchain technology and the business ecosystem can lead one to overlook some obvious realities, which includes the high cost of maintenance, the slow development of the technology, the high complexity of incentive design and the risks associated with spontaneous smart contracts. Our expectations of blockchain for the future of business and society are based on the long-term development of this technology. However, we are currently looking at the risks to the social and economic order that come with a fully decentralized technology.

These are our preliminary studies on the institutional economics of blockchain. The possible implications of blockchain technology in terms of social institutions as well as in the economic sphere are understood through discussions on topics such as distributed ledgers, dual governance structures and the institutional economics of cryptography. The key to this is that decentralized technology needs to be understood as not necessarily leading to fully decentralized social institutions and contracts. A more realistic social logic needs to be adopted. On the other hand, the social risks associated with a total reliance on an automatic technological contract need to be seen, and an over-emphasis on techno-supremacism needs to be guarded against.

In summary, the discussion of institutional economics in the digital age is divided into three chapters, which include economic and social institutional change, smart contracts and Coase Theorem, and institutional change and the digital economy. An understanding of the relationship between technology, society and the economy is of paramount importance. Changes in the way people work together are driven by technology to enable changes in the division of labour in society to be driven. This leads to the evolution of social institutions. In Rothbard's "Man, Economy and State with Power and Market", all references to production and exchange are based on an understanding of "man". At the same time, the understanding of institutions has to be based on the understanding of human behaviour in the digital economy. In the digital economy, technological supremacy is not the truth either. It is only when the balance between security and economy, and the inner logic of the institutional structure is grasped that the most appropriate institutional arrangements for the development of business and society can be found in the digital economy.

While the benefits of blockchain's institutional economics are clear, there are also those that need to be seen. For example, crypto banks with the transnational aspects [illegible]

Chapter 13
Money Supply and Demand and the Ecology of the Token Economy

After the institutional theory of digital economics has been explored, the topic related to monetary theory is once again discussed. Although the monetary theoretical ideas of digital economics have been initially introduced previously, and the basic ideas of traditional monetary theory and the monetary theory of Austrian economics have been sorted out, the in-depth study of monetary theory needs to be continued for three reasons.

Firstly, money is not only an infrastructure of modern economies and societies, but it is also one of the most controversial topics in economics. Both the idea of the neutrality of money and the theory underlying inflation need to be revisited.

Secondly, monetary theory in the traditional economic sphere has the disadvantage of being theoretically inadequate in the digital economy. It is particularly evident in the exploration of topics such as the supply and demand for money, the structure of prices and economic cycles. It needs to be reorganized in the field of the digital economy.

Finally, according to the idea of token economy design of blockchain technology, analysis based on the study of monetary and banking systems is needed in the field of digital economy. At the same time, a new institutional arrangement needs to be explored based on the theory of economic eco-design.

For these reasons, the exploration of the topics in this chapter is divided into three parts, which include money supply and demand and the credit medium, the principles of economic eco-design, and the business logic of the token economy.

Money Supply and Demand and the Credit Medium

The topic of monetary relations and the credit medium needs to be explored first and foremost, with reference to the study of the nature of money. As we have discussed, the theory of Austrian economics emphasises many features that are neglected in mainstream economics, such as a completely free economic order and a theory of

Z. Liu, *Principles of Digital Economics*, Contributions to Economics,
https://doi.org/10.1007/978-981-16-9020-4_13

money based on an evolutionary order. However, these features are of great value in the digital economy. The digital economy has developed in such a way that disruption of the market by government or other institutions does not exist. Therefore, the exploration of monetary order in Austrian economic theory also has relevance in the framework of the digital economy. However, it is also important to learn from its strengths as well as to avoid its weaknesses in order to avoid a mere mechanical appropriation of the theory. We should discuss it in a practical context. It is discussed in three parts, which includes the value of monetary relations, the issuance of credit media and the expansion of bank credit.

The theory relating to monetary relations was first explored, which refers to the theory of the relationship between supply and demand for money. To be precise, the monetary relationship is defined as the rate of exchange between goods and services. When the monetary relationship is constant, there is no inflationary (expansionary) or deflationary (contractionary) pressure. The causes of inflation are analyzed here with emphasis. This is because it is an important contribution to monetary theory by Austrian economists. Mises argues that the result of government credit expansion is usually inflation. At the same time, he proposes the following three dimensions of analysis.

First, inflationary policies inevitably lead to excessive consumption and misinvestment, the essence of which is the waste of capital, ultimately to the detriment of the satisfaction of future desires.

Secondly, the need to adjust production and allocate resources is not eliminated by inflation. It will only slow down the process and increase the difficulties of allocation.

Finally, inflation is unsustainable and results in economic cyclicality as well as leading to the collapse of the monetary system. That is, government-issued money is usually used as a credit medium. And the increase in the credit medium is used by the government to do the granting of credit expansion, which is the underlying logic of inflation.

Then, the topic of credit medium issuance is explored, which refers to the discussion of the utility of money. The difference between money and ordinary production goods needs to be emphasised here. It means that for ordinary production goods, an increase in their supply increases the supply of consumer goods. And if the supply of other goods does not fall, the increase in the supply of consumer goods will contribute to the increase in the general utility of society. However, money cannot be used directly in the production process as a productive good. Money is always present in the cash balance of an individual. It cannot be used directly, but only for exchange. Therefore, when money is low, the exchange value of the unit of money increases. When there is more money, the exchange value of the monetary unit decreases. Based on David Hume's theory, if everyone wakes up and finds that their monetary assets have become twice as large, the wealth of society has not increased twice as much. Only the monetary unit has been changed, and it has been diluted. In short, any supply of money is being used to the maximum extent possible. Therefore, any social utility does not arise from an increase in the supply of money.

Based on the above theories, the opposition of Austrian economists to economic ideas such as the banking school and the token school is understood. In these theories, it is broadly believed that the market can be positively steered and controlled by means of interest rate restrictions or active credit expansion. In Rothbard's "America's Great Depression [36]", the business cycle theory of Austrian economics is mentioned. In fact, the expansion of money not only raises prices, but also lowers interest rates. And the rise in consumer prices is only the manifestation. The damage caused to the structure of capital is the main disadvantage of monetary expansion. Wrong signals are released by artificially low interest rates in order to make projects that would otherwise not seem profitable be noticed and invested in by entrepreneurs, ultimately leading to inefficient capital and a general rise in prices. In this process, only the credit medium is increased. And the input of means of production and labour of new firms inevitably comes at the expense of the means of production and labour of other firms, eventually allowing a bubble economy of false prosperity to emerge.

In short, the utility of money indicates an accommodative monetary policy of the central bank, which initiates over-investment in the capital goods sector and diverts investment away from the consumer goods sector. Rothbard said that the process of production is always longer than the process of money circulation. Prices rise before the production process is complete, which inevitably raises costs and lowers profits for businesses. As a result, once the banks had taken steps to stop the expansion of credit in order to prevent the collapse of the monetary system, there would have been a reduction in the scale of production and a mass collapse of businesses.

Finally, the relationship between the expansion of bank credit and the quantity of money is discussed, with its reference to fundamental thinking about the central banking system. This is informative for the understanding of mechanism design in the token economy.

At present, banks are the mainstay of the monetary system. The two most important mechanisms in modern banking are the "central bank" and the "fractional reserve banking system". As a result of these two mechanisms, the principle that money is essentially gold has been violated. The commercial banks, guided by central banking policy, intervene in the market through the expansion of credit in order to bring about the phenomenon we have just discussed. Of course, we are not arguing here that the central bank's motives for intervention are negative. Rather, it is because central banks cannot find the most favourable monetary policy at every moment. Let us take the example of the continued recession of the Japanese economy over the past two decades. Even though a prolonged policy of monetary expansion and low interest rates was adopted by it, this was seemingly ineffective in helping the real economy. Therefore, many Austrian scholars believe that the order of the free market can be achieved through the establishment of a 100% reserve system, the abolition of central banking mechanisms and the privatization of credit money.

The critique of modern accounting principles by the Austrian economist De Soto needs to be added here, with particular reference to the International Accounting Standards (IAS). He argues that the incorporation of IAS into the laws of different countries meant that traditional prudential principles were abandoned in order to allow the emergence of a widespread exaggeration of book values by stakeholders. As a result, a cycle based on accounting books was created in a speculative bubble. It means that the growing market value of a stock enters the books, but the elements that favour the growth of the stock are found in the books to further drive the share price. De Soto argues that the new international accounting standards will assess the valuation of companies based on more subjective criteria than traditional accounting standards, so that the creation of accounting values out of thin air will be condoned, ultimately leading to increased risk in the financial markets.

Based on the above phenomena, proactive changes in the accounting industry are being driven by the Big Four accounting firms through blockchain technology. In April 2018, the Blockchain Analyzer was launched by Ernst & Young to help Ernst & Young's audit team review transactions on the blockchain and to enable the development of automated audits of blockchain assets, liabilities and smart contracts. In 2016, the 'Digital Book Services' initiative was launched by KPMG to enable financial services companies to investigate blockchain applications. In December 2017, Bitcoin began to be accepted by PwC. And in April 2018, its first widely released blockchain audit service was announced. 2014 saw the launch of a "blockchain one-stop software platform". In August 2018, it partnered with MAS to facilitate the creation of a "voucher settlement" service by Singapore's central bank and SGX to enable the settlement of tokenized assets on different blockchain platforms. These are not only applications of blockchain in the accounting sector, but are also manifestations of the digital economy gradually changing the existing financial rules.

In the light of the above analysis, the order of the free economic market, the opposition to the central banking system and the critique of modern accounting standards espoused by Austrian scholars can be understood. In the field of the digital economy, here are three lessons that these theories teach us.

Firstly, its most fundamental value is the use and exchange value of money. So a token ecosystem based on exchange value needs to be built up in the digital economy ecosystem.

Secondly, damage can occur as a result of artificial inflation. Therefore, the order of the free economic market needs to be relied upon to compete for the right to issue currency. In the digital economy ecosystem, the value that each of the different ecosystem tokens possesses depends on the real products and services of the ecosystem behind them.

Finally, a different business logic from traditional accounting standards is provided by the mechanism of the distributed ledger. Overvaluation of networked business organizations can be effectively avoided.

The Principles of Economic Eco-design

After the demand for money and the value of credit have been explored, the principles of economic eco-design are explored. Here we hope that this chapter of the study will provide the reader with three benefits.

First, the underlying logic of the redistribution of capital factors brought about by money needs to be understood and sorted out so that the global changes in financial capitalism being experienced by the entire human economic system can be understood.

Secondly, the principles of digital business ecosystem design are further grasped through the understanding of traditional economic ecological design in order to enable the basic logic of digital ecosystems and how to build good economic systems in a digital economic ecology to be understood.

Finally, this chapter allows us to build the theoretical basis for a token-based economic ecosystem in blockchain networks, which refers to the problem of designing a token economy for blockchain that many researchers have focused on.

This topic is discussed from three perspectives, which includes the system of division of labour and development of industries, the system of prices and allocation of resources and the system of distribution of income and wealth.

The system of division of labour and development of industries was first explored, with the system of division of labour meaning the structure of industrial value chains or value chain systems. We have discussed the value theory of digital economics. When reflected in economic ecology, it corresponds to an overall assessment of the value of an industry. Currently, the system of accounting for national income in economics or the system of measuring gross domestic product (GDP) can only be used to measure part of the value of the economy, and many other related values cannot be measured. Specifically, value is measured in terms of the totality of material, social and spiritual values in a digital economy. Material value refers to the value that can be captured by accounting indicators such as the overall revenue of a business, and the value module that is most easily understood and analyzed. Social value refers to the ability of the digital economy to provide a better service and experience for the users of the sector, to build a complete and healthy ecosystem, and to address the real needs of the users. This component is directly related to the industry in which the ecosystem is located and the characteristics of the users. Spiritual values refer to the community consensus and culture that this digital economy ecosystem imparts. For the users, life is given value and meaning. For example, many community products have a high degree of spiritual value.

It is then explored from the price system and the resource allocation system. In the business ecology, the concept of value chain is important. In fact, the price system and resource allocation system of the entire industrial ecology depends on the players in the value chain. The study of value chains in modern business theory is divided into two parts, which include the integration of value chains within industries or firms. It corresponds to a deepening and refinement of the division of labour within the industry.

The change in the value chain of an industry is the most important role of technological progress and technological innovation in order to enable the hierarchical structure of the enterprise or organization to be changed, and ultimately the vertical and horizontal changes in the organization of the enterprise to be driven. Another part is the study of the ecology of the value chain, which enables the ecological efficiency of the chain to be improved. In other words, the efficiency of certain links is enhanced through the promotion of technology, so that the overall efficiency of the industry's value theory is driven up. In short, the reorganization of the price system and resource allocation system is driven by the reshaping of the value chain, which is one of the most important objectives and tasks of the blockchain network organization.

Finally, it is discussed in terms of the income and wealth distribution system. Business organizations or participants in the ecology at the higher end of the value chain not only have the power to speak and decide for the entire network, but also have the power to set prices, allocate resources and distribute income in the digital economy ecology. In traditional organizational ecologies, the price structure and distribution system of the entire chain is dominated by companies at the top end of the value theory. Increased competitiveness is driven through contractual subcontracting, and their ability to maximize the price of products and services that undercut other competitors in the value theory in order to gain excess profits. This is the analysis from an ecological perspective. When this logic is placed in the context of competition in the ecology of the digital economy, ecological competition in the digital economy can be understood as monopolistic competition style. There is inevitably only one winner in a single field, and ecological diffusion occurs.

On the other hand, the distribution logic of the entire business ecology has been changed by the blockchain network organization, which refers to the change in the corporate philosophy of the industrial-era business organization centred on maximizing shareholder value (maximizing market capitalization). The revolution in the internal governance structure of the modern corporation emerged as a result of the industrial revolution and modern industrial capitalism, which included the financial revolution in capital markets, the separation of corporate ownership and management, the rise of professional managerial systems and the hierarchical and bureaucratic internal organizational structure of the corporation. The business ecology of managerial goal maximization emerged as a result of the highly centralized model of corporate management and control of the industrial economy. Managers' objectives are based on their own pecuniary interests, the interests of their employees, the interests of their shareholders and the demands of their customers. In the digital economy ecology, it takes a new twist. For example, traditional equity incentives are abandoned by the consensus mechanism in blockchain networks. Equity requires the involvement of financial institution intermediaries and with stakeholders who are investors rather than consumers. As such, it is indirect rather than direct to the business's own ecology.

In blockchain networks, the distribution of income is rearranged through the token mechanism. It is not only a logic with breakthrough value, but it transforms the logic of wealth and income distribution in the entire industrial economy. The current risk and challenge is to facilitate its legal and compliant growth in the digital economy ecosystem.

These are my studies on topics related to the design of economic ecology. This topic has been discussed from three perspectives, which include the system of division of labour and development of industries, the system of prices and allocation of resources and the system of These are analyzed from both macro and micro perspectives so that the differences between the ecological design concept of the digital economy and the traditional ecological design concept of the industrial economy can be understood more thoroughly.

The Business Logic of the Token Economy

In previous articles, parts of token economy related topics have been discussed, which were mainly based on the idea of the application of tokens as a generalized money theory in digital economics. In this chapter, the content of this chapter and the practical application of blockchain technology are combined to allow the design ideas of the token economy to be explored. It refers to how the token economy is understood through the dimension of the support it provides to the real world and economic operations. According to this dimension, the following three aspects are explored, which include the cross-border idea of the token economy, the token economy and monetary theory, and the design idea of the token economy.

The application of the cross-border idea of the token economy was first explored in terms of how the token economy can actually contribute to the optimization of the real economy or the traditional internet economy. Like the Austrian school of economics, the theory is difficult to realize in practice because of the impracticality of completely subverting the idea of the existing economic order. No country or government will ever relinquish its current control of the economy. At the same time, a completely free and orderly economy does not correspond to the reality of most countries. Therefore, the application of the blockchain token economy cannot be completely divorced from reality and consider a world of pure "digital survival". This is an idea that comes from the geeky open source movement. The token economy should be studied through a cross-border, hybrid and improved mindset. On the one hand, intermediaries should be removed by decentralized technology mechanisms, so that the overall efficiency of economic operations is increased and the transaction costs of businesses are reduced. The transformation of real economic activity is realized through the value network capabilities of blockchain technology. On the other hand, the existing application paradigm and competitive ecology of the digital economy (which mainly refers to internet applications) is transformed through a completely decentralized application, which refers to the DAPP model. Ultimately, the existing monopolistic competitive ecology can be disrupted in order to enable the development of the entire digital economy ecosystem to be driven.

In short, the boundaries of digital assets are maximized by blockchain technology. The evolution of the existing business ecosystem is facilitated by the liquidity of digital assets. On the one hand, digital proof is provided through blockchain technology, which guarantees the authenticity and objectivity of digital assets to the greatest extent possible. It refers to the provision of "algorithmic credit". On the other hand, the circulation of digital rights to corresponding products and services is provided through the circulation of the value of digital passwords, with the right of use at their core. The essence of this circulation is the reliance on existing products and services and the scenarios for the use of the different ecosystems offered by the relevant blockchain networks, which refers to "commercial credit". The value of these two components should not be confused and should not be understood simply in terms of the concept of "digital currency".

Next, the token economy and cryptocurrency theory are explored. The question of the relationship between the nature of money and tokens is the main concern here. The nature of money as a medium of exchange is first explored. In fact, it consists of two parts of demand. On the one hand, it is used for the needs of consumption and production. On the other hand, there is the demand for it as a medium of exchange. Therefore, the exchange value of a medium of exchange (which refers to the purchasing power of money) should be the sum of these two components of demand. Mises uses the law of regression to show that the emergence of the demand for money as a demand for possessions for industrial purposes is influenced by its exchange value. Thus, the purchasing power of money, like all other goods and services, depends on demand and supply. In the traditional monetary theory of the quantity of money, the growth of money is considered to automatically raise the price level by a corresponding percentage without affecting relative prices. Whereas, in fact, the purchasing power of a unit of money is reduced by an increase in the quantity of money. At the same time, the relative income and price structure will inevitably be altered.

Based on this logic, here are three tokens that are of a nature relevant to the theory of money.

First, supply and demand are determined by the value of the currency. Supply and demand in the digital ecosystem determines the value of tokens as consumption and production, and the value of passes as exchange. It would therefore be unrealistic for tokens to be simply equated with digital currencies or even legal tender. Because in real economies, most tokens do not generate real demand and do not have the same value as a legal tender. We need to think about how to make token gradually become a credit medium with use and exchange value within the blockchain network, and through this way change the basic logic and boundaries of the traditional enterprise organization ecology, rather than through its equity properties for rapid securitisation of digital assets, which defeats the original purpose of the whole ecological design.

Secondly, the utility of money derives from the benefits and expectations that individuals have of acquiring its value. One of the most important properties of money is its liquidity. Therefore, the concept of 'neutral money' or 'money with stable purchasing power' does not exist. In other words, a dynamic theory of money should replace a static theory of money. Not only is money not neutral, but it is also not purchasing power stable. Therefore, Mises advocates the adoption of a free banking system in order to achieve a full 100% reserve system. In token economic theory, on the other hand, the disadvantages of a liberal economic order are to be promoted through monetary policy. How to make the token gain its use value and exchange value is key, and its refers to the creation of scenarios for the use of the token. On the other hand, how the value of the token is linked to goods and services has to be considered. This is the key to making tokens circulating within the network.

Finally, the purchasing power determines the utility of money. The operation of the market tends to determine the final situation of the purchasing power of money at the height when the supply and demand for money are in agreement. As a result, there is never too much or not enough money. The benefits of indirect exchange and use of money are fully enjoyed by everyone and anyone, irrespective of the size of the total amount of money. In other words, the redistribution of social wealth occurs as a result of changes in the purchasing power of money. In an economic system, the amount of money available is always sufficient for everyone to have access to all that money has to offer and can offer. For the token circulation mechanism in the token economy, the following is the sense in which the basic logic of such a monetary principle is understood. Firstly, the methodological significance of pursuing a stable monetary mechanism is not great, especially since intervening in the market through active monetary policy will lead to inflation. Secondly, the key to monetary policy is who drives the fluctuations in the value of money, the managers of the network organizations that set monetary policy or the free choice of each individual in the network. In principle, a freely fluctuating monetary order should be established as far as possible. Managers should provide the infrastructure for an ecological flow of value rather than implement overly interventionist policies.

The discussion on the topic of STOs (securities pass-through issuance) in the blockchain space needs to be added to, and its part of an important topic that has been discussed recently in the digital finance space. At the same time, it is an important perspective for understanding digital assets based on the logic of token. This topic has sparked interest not only from those in the blockchain industry, but also from those in the traditional finance industry. Unlike Bitcoin's attempts at "digital gold", STO offers a path to transform traditional finance by turning "securities" into digital tokens, and it also refers to the existence of Securities Tokens. Securities Tokens represent digital tokens of ownership of underlying assets or securities that are backed by real securities, such as public equity, private equity, real estate and trading funds. According to many industry players, most digital assets in the crypto economy are expected to become Securities Tokens by 2025. Many countries are also experimenting with innovation in this area. For example, the ASX is offering tokenized securities to the Australian public stock market by creating its own private chain.

The token we discussed before refers more to a functional token, which refers to a usage-based token economy. STOs, on the other hand, are securities tokens, which represent ownership, dividend rights, voting rights, etc. of a blockchain organization. Based on current US and European standards, most tokens have the potential to be recognized as securities, which refers to the mechanism by which a token gradually becomes a digital asset placed in the financial system. With the premise of being recognized as a security, blockchain organizations would be able to achieve fundraising through a securities registration scheme in order to be regulated by financial institutions. Based on the views of experts in the field of digital finance, an important reason for the widespread acceptance of STOs by blockchain organizations is the huge regulatory costs as well as the competitive disadvantages that are very much faced if the STO approach is not taken.

In fact, we see that the greatest value of a security token is that it is "programmable". This is the most essential property of digital assets. Through the programmability of digital assets, algorithms and software can gradually replace the role of human beings in order to perform simple operations ranging from the payment of dividends to the design and trading of complex securities (such as convertible bonds and other financial derivatives). Securities tokens are based on smart contracts. As a result, such transactions are executed completely automatically. On the one hand, the efficiency of the entire transaction is increased. On the other hand, the regulation of securities finance is thus made proactive in order to eliminate the transaction costs associated with complex legal institutional arrangements. Of course, the tokenization of securities also has the obvious benefit of making "money never sleeps" a reality. Digital securities are used to replace traditional securities. As a result, blockchain-based transactions can be traded and settled 24/7, meaning that real-time trading will become a real reality, replacing the T + 1 or T + 2 settlement mechanisms that now dominate the financial markets.

Despite all these benefits, we need to see the phenomenon that securities token is a more realistic option for markets where the financial system has developed over a long period of time and where financial regulation is more mature. It is still essentially a security, so the core of all regulation of STO is still the balance between the efficiency of corporate finance and the protection of investors' interests. According to the US SEC (Securities and Exchange Commission) policy, it will not change the existing definition of securities for the sake of, for example, securities crowdfunding or private placements. Therefore, STO pass the Howey Test to verify their securities attributes. If it passes, then it can become a standard for securities. For our relatively lagging and more stringent financial regulatory system, it may be difficult for STO to be recognized as a formal form of securities anytime soon. This is because any financial innovation, whether or not it is technologically transformative, is dependent on the accompanying institutional arrangements. This is the basic logic behind our understanding of the future trend of STO.

Finally, the design ideas of the token economy are explored. Based on the above, some of the most important ideas about the design thinking of the token economy can be seen, which include the following.

First, there is a close connection between the idea of token value and the theory of money. Without its direct relationship with legal tender being considered, the ecology of token needs to be considered based on the theory of exchange value. A usage-based token would change not only the property rights basis of the ecology, but also the mechanism of income distribution throughout the ecology. This is one of the most important innovations in blockchain networks. It follows from the cases in the history of the development of the digital economy that the development of

Internet-based community forms and technologies has led to a consensus on open organizations. It is based not only on the borderless nature of the space in which information is transmitted on the Internet, but also on the increased efficiency of real-time communication and coordination between groups as the immediacy of information transmission increases. But so far, the incentive problem of open organizations has remained unresolved. The token economy is an important solution to this part of the problem, which refers to how the interests of all parties in a blockchain network organization can be arranged through incentive-compatible design.

Secondly, stimulating growth through monetary policy is certainly not the way to go. What is truly valuable is the value that tokens gain through the construction of value within the ecosystem. Like money, tokens are not used directly for consumption and production, but for the exchange process within it. It is therefore centred on how to create more scenarios to facilitate exchange, rather than trying to attract users through the 'equity' properties of tokens. Based on mechanism design theory, every rational person has a self-interest in a market economy. Therefore, a good institutional arrangement needs to match the pursuit of individual interests with the collective values that the enterprise needs to achieve, and to promote changes in production relations by maximizing the fit between individual interests and the interests of the enterprise. Obviously, over-stimulating economic demand at the expense of the value aspirations of the long-term development of the enterprise behind it. This does not satisfy such conditions and is not a rational institutional design arrangement.

Thirdly, the distribution, governance and active management of the value of tokens in a token economy are all manifestations of a series of interventionist policies. Austrian scholars would argue that in an ecology under direct free economy elections, managers are only responsible for managing goods, labour and basic contracts, not the value of tokens. The digital economy ecology undoubtedly has greater freedom, and it needs to find the right balance between efficiency and fairness in its governance system. We have seen that innovation in the digital economy often starts with an open source project, and most of these projects are innovations without a clear production goal and without efficiency. Blockchain technology, on the other hand, attempts to motivate community participants by clearly defining the benefits for all parties involved through incentive-compatible algorithmic rules and institutional arrangements, ultimately enabling distributed organizations to collaborate efficiently in an ungoverned state.

The above are some thoughts on the design of the token economy based on monetary theory and economic ecology design theory. On the one hand, the differences between legal tender and token need to be understood, and the differences are determined by the linkages and differences between the digital economy and the real economic ecology. We cannot just focus on the benefits of technological innovation and ignore the risks it entails. As we have said before, security and risk are two factors that need to be taken into account in all institutional arrangements.

On the other hand, the theoretical value of monetary policy-related references to the token economy needs to be understood. Not only is this key to our understanding of the entire digital economy, but it is also an important area of research in the future trends we are now seeing regarding the institutional dividends of digital technologies. Token is evolving in content and value, and we need to iterate on this part to adapt to the rapid changes in the digital economy ecosystem. At the same time, we also believe that innovation needs time cycles so that it can be practiced and continuously improved, eventually adapting it to the times.

Chapter 14
Capitalism from a Digital Economics Perspective

After the monetary theory of the digital economy and the institutional economics have been discussed, the issue of capital in the digital economy needs to be explored. Here are three main reasons why the logic of capital is being discussed.

Firstly, monetary credit has a very important link to capital, which is mainly manifested in the global financial capitalist economic environment that we are currently facing. This is the context not only for the emergence of blockchain technology, but also for the entire business enlightenment movement of the digital economy. The current state of development of global capitalism and the new ideas of capitalism in the digital economy need to be analyzed in terms of their historical origins and academic logic.

Secondly, the concept of rent, which was once discussed, actually involves the concept of social capital in the economic paradigm of complexity. The theory of social capital and the understanding of heterogeneous capital is a very important dimension. The idea of homogenized capital and the expansion of traditional financial capital has not only led to the emergence of financial crises, but also to a series of problems such as the inequality of global income distribution and the blind expansion of monetary credit. This area is explored in depth in the theories of the Austrian school of economics. At the same time, it has a more explanatory power with regard to the perception of capital and money.

Finally, the question of the structure of capital will be covered by the discussion of innovation in network organizations, which is inevitable. The understanding of the concept of structural holes contributes to the understanding of the fundamentals of network innovation as a whole, and the foundations of the innovation theory that has been discussed. On this concept, not only the basic framework of a theory of capital based on the digital economy is to be established, but also the basic logic of the understanding of innovation and economic development in the digital economy.

In the light of the above judgement, the issue of capitalism from a digital economy perspective will be discussed from three perspectives, which include Dispute of Two Cambridge and Complexity Paradigm, Capitalism of the Austrian School of Economics, and Capitalism and Innovation in the Digital Economy.

Z. Liu, *Principles of Digital Economics*, Contributions to Economics,
https://doi.org/10.1007/978-981-16-9020-4_14

Dispute of Two Cambridge and Complexity Paradigm

There is a famous case in the history of economics, known as the "Dispute of Two Cambridge". It belongs to the debate between the neo-classical synthesis of Post Keynesian economics and the neo-Cambridge school on the main issues of Keynesianism. This event is not only about important debates and transformations in the history of economics, but also about the relationship between dominant and heterogeneous capital that we are discussing. In particular, how the structure of capital is understood in terms of heterogeneous capital and how real-world economics is understood through a complex economic paradigm, rather than a neoclassical one. The history of Dispute of Two Cambridge refers to the major debates about the history of economic thought on the theory of capital and the theory of income distribution. Although many people are not aware of it now, it is still necessary for researchers in economics to understand the logic behind the evolution of this history.

The "Dispute of Two Cambridge" (also known as the "Cambridge Capital Controversy") refers to the neo-classical synthesis represented by the American MIT professors Samuelson, Tobin, and Soto and the "Cambridge Capital Controversy" (also known as the "Cambridge Capital Controversy") refers to the debate between the neo-classical school of synthesis, represented by MIT professors Samuelson, Tobin and Soto, and the neo-Cambridge school, represented by Cambridge professors Joan Robinson, Kaldor and Sraffa and the Italian scholar Passinetti. Although the first professors are better known in China (especially Samuelson), in fact the latter professors and scholars are no less accomplished than the first ones. Joan Robinson, for example, is the only woman among the world's leading economists. She not only created the theory of monopolistic competition, as well as developing and complementing Keynesian theory, but also created the new theory of capital accumulation. She is a leading figure in the Keynesian school of thought. She is also considered to be one of the greatest economists of our time.

In Dispute of Two Cambridge, Keynesian economics and neoclassical economic theory are combined by the former. The decisive role of technological relations in economic variables is emphasised. Its central idea refers to the fact that the supply and demand of economic actors who maximize their goals are determined by the pricing of goods and the pricing system of factors of production. At the same time, this is at the heart of neoclassical economics. The tradition of social relations analysis of the classical surplus economy is carried on by the latter. The ultimate role of socio-economic relations (ownership and economic institutions) in economic variables is emphasised. At the same time, a theoretical system based on objective value theory and centred on distributive theory and based on which the formulation of social institutions was explored attempted to be established in order to change the paradox of capitalism's regulation of employment and inflation through the distributive system. This is also known as the New Cambridge School.

In the battle between the two schools, both consider themselves to be representatives of the Economics of Keynes. This is simply because the two schools have fundamental differences, which are reflected in two main areas.

Firstly, the general equilibrium concept of Marshall and Vallas was inherited by the neoclassical synthesis represented by Samuelson. The equilibrium of the economic system and its stability were emphasised and the capitalist economy was considered to be a "system of general equilibrium". Equilibrium theory can be understood as the basic analytical approach of the neoclassical school of economics. The Neo-Cambridge school, on the other hand, considers that the main argument of Keynes' "general theory" is to break the fetters of equilibrium and to take into account the characteristics of real life. It means that "the past cannot be recalled and the future cannot be known with certainty". In other words, real-life behaviour is subject to uncertainty in terms of lack of information, and economic life is largely dealt with on the basis of accepted management. From a methodological point of view, the uncertainty, the complexity of reality and the historical system of capitalism in economic analysis are emphasised by the New Cambridge School, which is closer to reality than the neoclassical synthesis, rather than focusing on aesthetics as in the general equilibrium system.

Secondly, the micro-analysis of factor supply and markets was used by the neoclassical school to fill in the gaps in Keynesian economics, which led to the emergence of a strange "macro-micro" theory. This is why the Cambridge school of economics considered the neoclassical school to be a "bastard" economic theory. This is because the approach has reverted to the pre-Keynesian tradition and the integrity of the Keynesian system has been completely destroyed. The New Cambridge School, on the other hand, argued that Keynes lacked a theory of value and distribution. Therefore, an economics based on an objective theory of value needed to be constructed back in the theories of classical economists such as Ricardo. In short, the New Cambridge School not only pointed out the problem of unequal distribution of income in capitalist societies, but also focused on the distribution of income in the process of economic growth. This is an important reason why we are now seeing the conservatization of the world economy and the ebbing of globalization. The methodological error of the neoclassical synthesis in its theoretical system is that it tries to confuse the "equilibrium model" or "equilibrium view" with the "historical model" or "historical view". "The Neo-Cambridgers also point out that in reality this is a very different approach". The Neo-Cambridgers also pointed out that in fact these two opposing methodological concepts are incompatible with each other.

Finally, the neoclassical economic synthesis refers to physical capital from the point of view of the concept of capital. However, it does not consider the measurement of capital used in the production function. The New Cambridge School argues that there are multiple measures of capital. The only one that works in theory is the one that is consistent with the New Cambridge value theory of the cost of labour wages plus a mark-up on profits to determine prices. In other words, when capital is seen as labour accumulated in the past, the amount of capital is calculated in terms of what Sraffa called "the amount of labour reduced to a period of time". But then the productive nature of capital is in fact a reflection of the productive nature of labour, and this needs to be recognized. Only labour creates output and wealth, while capital is merely an aid to production accumulated by past labour. Thus, the productivity of capital and the "marginal productivity of capital" do not exist. Therefore, under the only viable measure of capital, the neoclassical synthesis of the "production function" and the theory of distribution and growth would be bankrupt.

This is our review of the "Dispute of Two Cambridge", an important polemic in the history of economics. Sadly, although the Cambridge School did in fact win the polemic, their theories were never mentioned again after the death of several scholars. They won the debate but lost the discourse in economics. What is notable, however, is that the real economy did not work as the winners expected. Neoclassical economics' explanation of the real world has also been greatly challenged. This is the basic reason for our research and discussion in the field of the digital economy. That is, a framework for economic solutions that is closer to the real economic world and that is specific to the unique and innovative economic field of the digital economy is sought to be established.

Capitalism of the Austrian School of Economics

Dispute of Two Cambridge is discussed, and we learn that neoclassical economics has dominated the field of economics. But a long-standing theoretical crisis has persisted and has never been completely resolved. However, it is important to note that the above-mentioned debate was a debate between the two main post-Keynesian schools of thought. The end result was the success of neoclassical economics in fact (the New Cambridge School won theoretically, but it disappeared in fact). However, either school belongs to the admirers of Keynesian theory. And Keynesian theory has become part of mainstream macroeconomics, and its an important basis for government intervention in the economy. We focus in this chapter on the Austrian theory of capital in economics, and in particular on the heterogeneous ideas about capital and interest, and the structure of capital, derived on the basis of human action.

First, we discuss the idea of a theory of capital based on human action, whose results are the work of Mises. For the actor, time is inseparable from human action, and it is impossible to conceive of an action that does not take place in time. An action can acquire greater subjective value in two ways: firstly, by enabling the actor to achieve results that are more valuable in his subjective view and that could not have been achieved by a shorter human action. The second is that it makes it easier for the actor to achieve more results than would be possible through the actions of a shorter person. In other words, the longer the process of human action lasts, the greater the value they achieve, and it is precisely a certain length of time that separates the actor from the goal. In this process, the actor develops a so-called "time preference", which means that each individual prefers to satisfy his or her own needs as much as possible, and that "current goods" always have more subjective value than "future goods". The "present good" always has a more subjective value than the "future good". In the intermediate stages of each action process, each action process can be defined as a "capital good", i.e. a factor of production that is subjectively paperless for each intermediate stage of a given action process. i.e. the subjective assessment by the actor of the market price of the capital good.

Secondly, the contribution of Bohm-Bawerk, a scholar of the Austrian school of economics, is discussed. He was not only a follower of Menger's theory, but also a good friend of another Austrian School economist, Wiesel. His most important book is "Capital and Interest". In it he reviews in detail the debate on interest, from the time of Aristotle through the supply of usury by the Church in the Middle Ages to the series of controversies in the late nineteenth century arising from Marx's theory of surplus value and the assertions of other socialists. Bohm-Bawerk sums up the former as "having the present in order to consider alternatives and the future" (Schumpeter) and the latter as "roundabout production is more efficient". The latter was summed up by Irving Fisher in his Theory of Interest. Fisher, in his "theory of interest", continued and developed his theory, summarizing the three factors as two psychological factors, namely "impatience with income" and one productivity factor, namely "opportunity or return on cost". In the course of his research, Bohm-Bawerk understood capital as "a group of products as a means of goods" and "a collection of intermediate goods", an idea that was then carried forward by the later Rahman to deal with "heterogeneous" This idea was then carried forward by the later Rahman to deal with theories related to "capital".

Thirdly, Rahman's contribution is discussed. He inherited some of Bohm-Bawerk's theoretical ideas. However, he argues that Bohm-Bawerk's model is essentially a macroeconomic model and that its does not provide an adequate basis for a genuine Austrian school theory of capital. In Rahman's "Capital and its Structure" [37], it is worth noting the following three basic ideas. These three ideas constitute important elements of the Austrian School's theory of capital.

Firstly, there are important flaws in the Bohm-Bawerk theory, particularly in the discussion of the "average length of the production cycle", which misrepresents the diversity of the economic world. Therefore, the capital theory of Austrian economics needs to be studied at a fundamental level, i.e. at the micro level of the formulation and execution of production plans. It is interesting to note that Bohm-Bawerk insists on using the concept of subsistence fund to discuss capital goods. That is, the expected sacrifice of these production processes is evaluated in terms of reduced direct consumption when choices are made between production processes of varying length. The concept of the sistence fund is not only central to Bohm-Bawerk's theory of capital, but is also the most fundamental Austrian school component of his thinking. And his concern with forward-looking and intertemporal decisions is compressed into this concept.

Secondly, the basic theory of the Austrian school of economics is the recognition of the diversity of goods and services, by which is meant the recognition of the heterogeneity of capital (the existence of heterogeneous capital). And the existence of heterogeneous capital is the basis of the theory of capital structure. Rahman argues that Bohm-Bawerk only applied this theory in order to propose it for the theory of interest, but in fact the core of the theory of capital structure is to enable the form, order and consistency of capital structures to be explained in terms of human action. The recognition of the existence of heterogeneous capital also means that the role of the entrepreneur is recognized. Because the entrepreneur determines the allocation of heterogeneous capital, the size of profit emerges in a stable equilibrium (in fact there is no such equilibrium). Profit itself is therefore a non-equilibrium gain obtained through the dynamic allocation of capital.

Finally, based on the model of capital heterogeneity, which refers to heterogeneous capital, we see that the market works like the following. In the commodity market, each firm with its own portfolio of capital is in disequilibrium and contributes to a constant adjustment of the capital structure by acting in this context. The particularity of each firm lies in the constantly different interpretations of knowledge flows, in the way in which different entrepreneurs find the right expression for their specific capital portfolios, which are constantly changing in structure, thus bringing about the impetus of new investments on the market. And the social capital structure is the synthesis of all these capitals, based on the competitive and orderly dynamics brought about by the market process theory, where capital goods that do not fit into the existing portfolio are eliminated and those that are suitable for development continue to evolve, where it is the entrepreneur's decisions rather than marginal utility that drive the restructuring of heterogeneous capital, thus adapting to changes in present and future expectations, changes in costs and benefits and changes in the market. This is the essence of the Austrian school of economic capitalism as derived from the theory of heterogeneous capital and the market process.

In summary, the Austrian school of economics capital theory refers to the relationship between the structure of social capital and capital mix by treating it as a macroeconomic quantity and a microeconomic quantity. In other words, entrepreneurs drive the evolution of the social capital structure through different combinations of capital. Keynesian capital theory, on the other hand, is based on the efficiency of marginal capital, which refers to treating all capital as homogeneous. The Keynesian view of GDP as the end and human behaviour (investment, consumption as well as exports) as the means is the natural derivation of the homogenized capital theory, according to which everyone is a slave and a means to economic growth and economics becomes a slave to the application of mathematics, with mainstream economists looking at how to drive economic data through more subtle models, forgetting that human behaviour and demand It is human behaviour and demand, especially differentiated demand, that is the goal of economic growth.

Paul R. Krugman once said that "economics moves in the direction of least mathematical resistance". This inertialized thinking has led to the collapse and anachronism of mainstream economics. The Austrian school of economics, on the other hand, starts with human behaviour, and its the most thoroughly understood theoretical system of market economics available. The construction of a theory of capital in digital economics, on the other hand, is built on the theoretical ideas of the Austrian school of economics. Particularly with regard to heterogeneous capital and the subjective value theory of time preferences, an important basis for our understanding of capital in the network economy is constituted.

Capitalism and Innovation in the Digital Economy

After the Austrian school of economics, in which the theory of capital centred on human action has been discussed, one of the ideological foundations of the capital theory of the digital economy is understood, with its reference to the concept of heterogeneous capital and the inner logic of structural hole networks. The development of social capital theory is next highlighted. On the one hand, the concept of social capital theory is one of the biggest foundational differences that define digital economics from other areas of capital theory, with its emphasis on the role of human interrelationships. On the other hand, the theory of structural holes developed from social capital theory is also an important theoretical basis for explaining the capital structure of the network economy. Therefore, the basic concepts and developments related to social capital will be presented. At the same time, its application to areas related to the network economy will be briefly presented.

In addition to this, we will present the relationship between financial capital and corporate technological innovation. The intersection of two topics, the virtual economy (finance) and the digital economy (innovation), is involved here. In the interests of space, we will focus here on the basic framework for the analysis of financial capital and technological innovation and answer the following three fundamental questions. First, how does financial capital incentivise firms in the digital economy to innovate. Second, how innovation in the digital economy plays out in the financial sphere. Third, how to understand the financial logic of corporate innovation. Once this part is understood, an intrinsic link between finance and technological innovation can be established. Ultimately, the theory of capital from a digital economy perspective will also be better understood.

Firstly, the qualities of the capitalist economic system are explored and can be summarized in roughly three ways. Firstly, the means of production on which economic activity depends are privately owned by individuals, and the system of private property rights is the cornerstone of the capitalist economic system. Secondly, the main motive of the productive activity of the enterprise is to maximize profit or shareholder value. Thirdly, the coordination of economic production and business activities relies primarily on price signals from competitive markets, particularly financial markets, and the financing of money and credit. In other words, money and credit creation are central features of the capitalist economic system, and we are currently in an era of global finance capitalism. According to economist Xiang Songzuo's book "The New Capitalism " [38], global financial capitalism has three qualities. Firstly, the contemporary human economic system is one in which private property rights and market competition are the basic rules of operation. Secondly, the contemporary human economic system is a globalized economic system, that is, a globalized allocation of resources, a globalized division of labour and a globalized capital market. Thirdly, the contemporary human economic system is a system dominated and dominated by financial markets and the prices of financial assets. Therefore, globalized financial capital is not only the most important economic factor, but also the source of globalized economic crises.

Global finance capitalism is based on industrial capitalism and commercial capitalism. While traditional economic forms and market exchange patterns are the starting point for the whole of economics, commercial capitalism is nothing more than the domination of resource allocation, income distribution and the price system of countries by commercial capitalists. Industrial capitalism, on the other hand, was dominated by industrial capital in terms of resource allocation, income distribution and the price system, and in the era of financial capital, it was financial capital that dominated commercial and industrial capital, and financial capitalists began to dominate several of these elements. In the era of global financial capital, multinational corporations and financial institutions became the dominant players in the global industrial division of labour system, resource allocation, price system and the macroeconomic policies of each country. This process began with the creation of the unanchored monetary system as a result of the collapse of the Bretton Woods system in 1971, as we mentioned earlier, and the underlying logic was the global liberalization of market policies leading to competition and the division of labour. Its main drivers are the complex financial products generated by modern finance and the development of information technology. This is not only the root cause of the global financial capitalism behind the global financial crisis to date in 2008, but also the starting point for the study of blockchain technology in our understanding of the whole wave of the digital economy.

The origins and development of the concept of social capital are then discussed. In 1980, the French sociologist Pierre Bourdieu published an article in the journal "Studies in the Social Sciences" entitled "Essays on Social Capital". In this article, the concept was formalized and defined as "a collection of actual or potential resources related to an enduring network of mutual tacit or adult relationships, which are more or less institutionalized". In 1988, Coleman published an article in the American Journal of Sociology entitled "The Role of Social Capital in Human Capital Creation". The concept of social capital was developed in the sense of social structure. On this basis, a theory of "economic sociology" was developed. In addition, Putnam, a professor of sociology at Harvard University, drew much attention to this concept. His book "Making Democracy Work", a study of Italian administrative government, has drawn much attention and discussion to the concept of social capital. The latter, in turn, has conducted an in-depth and systematic study of social capital based on political science. In short, the concept of social capital has been applied in different fields such as history, sociology, political science and economics. And we can understand social capital from the following three perspectives.

Firstly, this is the view put forward by Pierre Bourdieu from the point of view of fields and relations. He argues that domains are made up of different social elements that are connected, and that different elements of society exist and function in a field by occupying different positions. The field is like a social web, and positions can be seen as the links in the web, with members and social groups having access to different social resources and rights as a result of the different positions they occupy. "Bourdieu sees the field as a network of relations formed by various elements, a dynamic process of change, the driving force of which is social capital". Classifying capital into three types: economic, cultural and social capital, Bourdieu concentrates on the distinction between capital and its interactions. It is argued that capital is interchangeable with each other. Bourdieu suggests that social capital is "a collection of actual or potential resources that are inseparable from the possession of some enduring network. This network is familiar, recognized, and institutionalized, and is linked to the membership of a group. It provides support for each member in terms of collective possession of capital, providing credentials that earn them prestige." Social capital exists in the form of a network of relationships.

Secondly, this dimension is the result of research done by Coleman from the perspective of social structure. He argues that the very purpose of social capital research is to study social structure through the study of social capital. Coleman states: "Events that embody the interests of some actors are partly or wholly under the control of other actors. Actors engage in various exchanges with each other in order to realize their own interests... as a result, social relations are formed that persist." He called social structural resources as capital property owned by individuals social capital. Coleman incorporated the work of Granovot and Linnan and Bourdieu, mentioned below, into his theoretical framework and developed his theory of "social capital".

Finally, social capital is discussed from the perspective of social resources. This is the work of sociologist Linnan. Linnan first developed the theory of social resources. According to Linnan, resources are "things that are perceived as valuable by a society or group through certain processes, the possession of which increases the chances of survival of the possessor." He divides resources into personal resources and social resources. Personal resources refer to those resources that are at the disposal of the individual such as wealth, appliances, natural endowments, physique, knowledge, status, etc.; social resources refer to those resources that are embedded in the network of one's social relations, such as power, wealth, prestige, etc. Such resources exist in interpersonal relations and must be acquired through interaction with others. The use of social resources is an effective way for an individual to achieve his or her goals, and personal resources largely influence the social resources available to him or her. On the basis of the social resource theory, Linnan also proposed the social capital theory.

Social resources are only associated with social networks. Social capital is a social resource mobilized from a social network. According to Linnan, social capital is "a resource that is invested in social relationships and is expected to be rewarded in the marketplace, it is a resource that is embedded in the social fabric and can be acquired or moved through purposeful action". Linnan defines social capital by emphasising its pre-existing nature. It exists within a social structure and people must follow its rules in order to obtain the social capital they need to act. At the same time, the definition illustrates the dynamic nature of human action. People can acquire social capital through purposeful action.

In short, "social capital" is one of the three basic forms of capital (the other two include economic and cultural capital). It is a collection of actual or potential resources acquired through the appropriation of an "institutionalized network". This "institutionalized network" is linked to membership of a group, and the acquisition of this membership guarantees the individual "prestige" and consequently material or symbolic benefits. For a specific individual, the amount of social capital he possesses depends on two factors. One is the size of the network of contacts that the actor can effectively use. The other is the amount of various forms of capital held by each member of the network. If social capital is used properly, it will be highly productive. This is because it has a high capacity for self-enhancement.

In order to accumulate and maintain social capital, individuals must spend a considerable amount of time and energy without interruption. This is the only way in which simple, contingent social relations can become an "obligation". This combination of temporal characteristics and the fundamental ambiguity of social exchange itself makes the use of social capital a "subtle economy of time", and an understanding of social capital makes it possible to understand the logic of the way in which community-centred network organizations are constructed in digital economics, and the way in which network economies are formed with The structure of the "structural hole" and the "monopoly economy" at the heart of the network economy is how it is formed. Building an ecology of niche economies in blockchain networks with social capital at its core may be a key area of focus for our study of the ecology of the digital economy, and a key approach to understanding capitalism in the context of the digital economy.

Finally, the relationship between corporate innovation and financial capital in the digital economy is discussed. Innovations in the digital economy are usually innovations in open networks. As such, financial capital is able to act on their innovations through more complex systems. Specifically, it can be divided into three levels, which include the micro level, the meso level and the macro level. In the interests of space, we provide a broad overview of the underlying logic and an outlook on the direction of future research.

Micro-level innovation incentives usually refer to the influence of firm-level factors on the innovation mechanism, including external and internal factors. External factors usually refer to organizations or individuals whose main objective is to invest in and analyse firms, such as financial analysts, external investors (which usually include institutional investors, hedge funds, banks, etc.). Internal factors refer to the financial elements within the innovative firm, including the liquidity of the firm's shares, compensation and incentives, the firm's management team, supply chain finance, etc.

Most of the time for technology companies, the role of external incentives tends to be positive at the beginning of the organization's innovation and entrepreneurship (e.g. the role of early and mid-stage venture capital for innovative companies). It has a negative effect after the company goes public or becomes a large-scale enterprise (e.g. the innovation capacity of listed technology companies declines). This is not only because external factors tend to put pressure on a company's share price or market capitalization in the later stages, but also because the core of the post-IPO innovation team tends to move away from the specific research work of technological innovation. Firms, in turn, are often caught in the innovator's dilemma, unable to escape the external pressures brought on by capital. In contrast, in the course of research into factors internal to the firm, we see a greater role for people than for organizations. That is, it is more effective to motivate people to take more risks and innovate than to change the organizational structure of the firm. This also relates to the issue of institutional arrangements that we have discussed previously.

Meso-level innovation focuses on the impact of market factors on innovation in technology firms, such as competition in industrial markets, competition in the financial sector and corporate taxation arrangements. We have seen that industry competition is particularly intense in the digital economy. This is because monopolistic "winner takes all" competition is common in the digital economy, such as the fierce competition that has occurred in China's shared mobility and group-buying markets, where only one or two firms remain in a single industry and thousands of others die out in a short period of time. Financial capital has often played a very important role in this, and we have seen that the institutional investors behind the companies have often become an important force in the market competition, and their will has been carried out to a large extent in the fierce market competition. There have been two very important cases in the Chinese internet market. The first was in the shared mobility sector, where the final outcome of several major competitors was almost determined by the timeliness of the introduction of the power of capital and whether capital was fully committed to supporting the business. The other is the massive short-selling of US-listed Chinese stocks, which led to heavy losses for many of them and forced them to return to domestic listings. At the meso level, the influence of financial capital on technological innovation in the digital economy is very complex. Capital markets play a role in optimizing the allocation of resources, facilitating corporate innovation and economic growth. However, the excessive short-term pressure on firms caused by capital markets can also lead to negative externalities such as short-sightedness and speculative behaviour.

The impact of financial capital on firm innovation at the macro level is mainly reflected in the design and change of financial mechanisms at the national level. In particular, the impact of policy elements, legal elements, cultural elements and other institutional elements on firm innovation. The research work at this level is reflected in two main areas. The first is the study of the relationship between financial capital and firm technological innovation in emerging market countries, such as for developing countries, and the relationship between financial markets and technological innovation capabilities. The other is the study of the relationship between institutional innovation and corporate innovation, such as the relationship between industrial innovation mechanisms and SOE reform and innovation in China. In fact, research into these two directions involves the analysis of the capital theory of digital economics in the specific field of industrial innovation and development economics research. This is also the focus of our future research at the intersection of the two fields of innovation and finance.

The above is the scope of our research on the intersection of capital theory and innovation theory in digital economics. Based on the micro, meso and macro frameworks we have covered a number of research topics and projects. The theory of digital economics is still in its early stages. Therefore, research on these topics is being implemented step by step. We hope that the presentation of this chapter will stimulate interest in research in this field and establish an intrinsic logical framework for the relationship between innovation and capital theory in the digital economy. In the future, we will gradually exhibit more results in related fields.

innovate than to change the organizational structure of the firm. This also relates to the issue of institutional arrangements that we have discussed previously.

Chapter 15
Economic Order in Open Network

Once the foundations of the capital theory of digital economics have been explored, some elements of this theoretical framework need to be analyzed in depth, such as the structure of capital, the quantity of money and the subject of the price mechanism of transactions. Digital economics addresses the economic structure of networks. As such, it will be discussed on the basis of the logic of open networks, which is concerned with three main aspects.

Within the theme of trading networks and capital structure, the networks constituted by random trading behaviour in the digital economy system are first and foremost discussed primarily as the underlying logic for understanding the microstructure of markets in the digital economy. The impact of capital structure on the efficiency of the overall trading network is discussed not only in terms of the relationship between structural hole networks and trading efficiency, but also in terms of the relationship between the industrial and financial structure of technological innovation, which needs to be explored based on the logic of financial structure.

Secondly, how the ecology and models of money are theoretically analyzed in a digital network needs to be explored within the theme of the open economy and monetary ecology. In particular, how risks due to value uncertainty in a virtualized monetary ecology can be avoided. Here, the internal links between the entire monetary ecology and the open network need to be analyzed.

Finally, theoretical models related to market prices in complex networks are explored in the theme Token Economics and Price Theory, as well as the mechanisms of market price formation for Tokens of blockchain networks based on information theory and complex economics are analyzed. In particular, market price mechanisms based on social choice are analyzed.

Based on the above three chapters, the establishment of an economic order within an open digital ecological network will be systematically explored, with the main element being how the monetary order is established. At the same time, the logic of trading networks and technology creation from capital and financial structures is understood, as well as the open economic order of niche economies and the essence of the Token economy is explored.

Z. Liu, *Principles of Digital Economics*, Contributions to Economics,
https://doi.org/10.1007/978-981-16-9020-4_15

Transaction Network and Capital Structure

The theory of money based on exchange value has been explored before. And when the process of exchange is understood within the logic of the market, it denotes the act of trading by people and the different decisions made in the transaction. The British economist William Stanley Jevons once said, "In the beginning, a market was a public place where a town sold food or other goods for sale. Its meaning was later expanded and it was used to refer to a place where people with close commercial ties and large quantities of certain types of goods were traded". Indeed, when the digital economy developed, marketplace transactions no longer required a fixed place, but rather it had an ordered structure of people from different geographical areas trading randomly in a network. In this chapter, trading networks as well as capital structures will be explored. A theory of market processes based on complex systems and spontaneous order will be discussed here. Also, market trading mechanisms based on structural holes will be analyzed.

The mechanics of trading networks will be discussed first. Three points can be drawn from our study of value network theory and capital theory in relation to digital economics. First, the view of monetary neutrality has to be discarded. The relationship between economic variables and the value of money should be understood in the context of the theory of money supply and demand. Secondly, the role of money in the modern economic system is crucial, as it is a key element in the understanding of the trading mechanism and the ecosystem of the market. The price of money, like other commodities, reflects subjective values. Thirdly, the existence of market uncertainty leads to the mechanisms of the market being considered based on complex networks. Thus, the key to our understanding of the mechanisms of trading networks is heterogeneous capital. Murray Rothbard has said that "every voluntary trade in goods on this basis also leads to Pareto Optimality. Because it occurs only under the condition that both parties to the transaction are expected to make a profit, while the goods owned by others in the action remain unchanged."

In the digital economy, these qualities of transactions are also embodied by the organization of networks. In other words, in the network of transactions created by the blockchain or the Internet, the boundaries and efficiency of the network of transactions depend on the random choices of all individuals. What happens in a transaction is not an exchange of equivalence, but an active transaction in which each individual perceives that he or she is getting more value. This is a process of "subjective value" rather than an objective process of "marginal utility maximization". The price system determined by this random choice can therefore be deduced directly from monetary theory. For the sake of space, this process will be discussed briefly here. The premise of the "rational economic man" will no longer be assumed. Instead, the subjective value of each individual is chosen randomly in the network to form the price system. The formation of prices is no longer an equilibrium price ecology, it is a dynamic system that is described by probability statistics. And this system forms a "dissipative structure" in the information network of the digital economy, where information, energy and matter are constantly exchanged. The blockchain network organization ecology can be understood as a dynamic and complex system with self-organizing behaviour. In the market, the price signal (Token price) determines the state and evolutionary path of this network organizational ecology due to the ordered form of the system. The above is our understanding of trading networks in the digital economy.

The capital structure of the transaction will then be analyzed. The concept of social capital has been discussed before. On the basis of this theory, the concept of a structural hole is discussed, which refers to "a phenomenon in which one or some individuals in a social network are directly connected to some individuals, but not to others. There is no direct or intermittent relationship. From the point of view of the network as a whole, it appears that caves appear in the structure of the network". This leads to the concept of structural caves. From the point of view of social capital, on the other hand, structural holes can be understood as rent dissipation in the organization of the network, which is avoided to the greatest extent possible. In other words, networks are driven to replace the general mechanisms of ordinary markets through the mutual synergy of complex networks and the innovative capacity of entrepreneurs, in order to make the network organization of the digital ecology function as both a business and a market. The capital structure of systemic innovation is realized through money and information. In other words, the networking of capital can be realized through an ecology of network organizations. The capital structure in transactions is a reflection of the network structure mechanism of social capital. Entrepreneurial innovation, on the other hand, is valued through the recognition of the role of the structural hole. This is the basic logic of our understanding of the whole theory of capital structure.

For the study of the financial structure of the digital economy, some new ideas need to be put forward based on the ideas of the new structural economics [39] in order for the capital structure to be understood. Based on the neo-structuralist perspective, traditional finance argues that it is financial depth that determines economic performance. In contrast to it, neo-structural economics argues that it is the financial structure that determines economic development. The underlying logic is that financial services are embedded in the structure of industry, which in turn is determined by relative factor endowments. As a result, the following conclusions can be drawn. First, when the economy develops, the optimal financial structure will evolve towards a market-led approach. Second, the optimal financial structure that drives economic growth is different in different stages of economic development. Thirdly, for countries, the optimal financial structure is deeply intrinsically linked to the country's policies, laws and cultural values.

That said, the following similar conclusions can be drawn in the field of digital economics. First, the internal logic of financial structures depends on the industrial structure of the digital economy. The problem that most firms in the digital economy need to solve is how technological innovation is driven by innovation in the financial structure. Secondly, the question of how to structure finance differently at different stages is very important for companies in the digital economy, especially when the positive development of the company is driven by the capital structure. Thirdly, the question that companies in the digital economy need to think deeply about throughout the cycle of innovation is how to build a corporate culture and corporate system that drives the rational arrangement of financial structures. This is not only an important basis for understanding the capital theory of digital economics, but also an important application based on a neo-structuralist economic perspective.

Finally, the above theories are analyzed from the perspective of complex networks and information theory. Here, the methodologies of information technology theory, complex network theory and economics will be integrated and analyzed.

In the process of trading commodities, interacting individuals form markets. The complex system of "dissipation" described above is formed. In this system, price mechanisms are used to spontaneously form an orderly structure, a process of self-organization and evolution. The emergence of money as a unified measure of price signals is key to this process. At the same time, money is the universally accepted medium of exchange, which gives efficiency to people's transactions, thus enabling entrepreneurs to allocate resources more rationally and facilitating the division of labour in society. The market system is therefore a self-organized exchange behaviour. The analysis of this behaviour requires research based on a number of disciplines, including cognitive science, information science, psychology, behavioral science and complexity theory, rather than relying on mathematics alone.

Structural hole theory [40] needs to be understood as the emergence of information asymmetries that allow for the emergence of structural holes, which refer to the existence of opportunities for monopolistic competition. In the networks of the digital economy, there are just two mechanisms of competition. One is monopolistic competition, which is equivalent to imperfect competition. In this ecology, entrepreneurs drive the economy through innovation driven by the allocation of heterogeneous capital. The other is perfect competition, which refers to an ecology in which heterogeneous competition in a given field is transformed into homogeneous competition. It is a case of complete information symmetry. In monopolistic competition, the key is the personality of the entrepreneur and the ability to innovate, both of which are elements of monopoly formation. In perfect competition, on the other hand, the key is the ability to dominate and manage the ecology, which is the building block of a stable platform. In short, the structural hole theory is based on the way in which the competition for power represented by market resources in the traditional economy is replaced by a free competition. In a digital economy ecology, the key is to assemble the right ecology to create innovation opportunities for the players in the ecology in order to allow a perfectly competitive dynamic to develop.

Open Economy and Monetary Ecology

After the theories of market microstructure have been explored, the elements related to monetary theory mentioned in the previous lectures are combined to discuss the open economy and the value of money. This chapter covers the following three parts. The first part is the process of money virtualization, which refers to the analysis of the metal theory of money and the symbol theory of money. This is a discussion of the intrinsic value of money. The second part is the analysis of the quantity theory of money, which refers to the analysis of the quantity theory of money in an uncertain open market. The exploration of relevant theories of monetary policy in neoclassical economics needs to be presented. The third part deals with the basic components of price signals. That is, the intrinsic order of the monetary ecosystem can be determined by which key price signals.

The process of money virtualization was first explored. Throughout economic history there has been a debate between the metal theory of money and the symbolic theory of money. The metal theory of money means that money is considered to be a commodity and must have intrinsic value. Therefore, precious metals, or money minted from precious metals, have monetary value. Aristotle, Adam Smith, Ricardo and later Mises of the Austrian school of economics all inherited this view. In the symbolic theory of money, money was seen as a symbol of knowledge as a medium of exchange, which required only legal solvency and not an intrinsic commodity value. Plato, Thomas. Aquinas, and later the masters of the neoclassical school of economics inherited this view. Throughout the millennia of human economic history, money has in fact been progressively "demonetized" from the real. When the state was formed and the credit mechanism of banks was established, money was gradually separated from commodities in the process of buying and selling, so that it became a special general equivalent. In this process, money is used as a medium of circulation, a means of payment and a store of value, which has a positive effect. The economic cycle is affected by the expansion of credit and inflation, which is a negative effect.

The question of this process has been discussed many times. However, there is no doubt that this process of the continuous virtualization of money forms has evolved from the development of civilization. Thus, whether economists like it or not, the reality is that the role of money as a store of value and as a medium of credit derived from liquid assets has been exploited by the financial sector, which has expanded money exponentially in the form of money multipliers through the expansion mechanism of bank credit. In particular, the securitisation of assets and a series of financial innovations drove the circulation of money through the economy at tens of times its real volume. This led to a paradox in which fragile economic growth needed to be sustained by more money, while money in the form of excessive credit expansion kept the economy confined to an irrational boom from which it could not escape. The basis of all this is the symbolization of money. In short, the global financial leverage and non-linear expansion of credit brought about by the liberalization of the capital account is structured by the symbolization of money. As a result, national monetary policies have gradually lost their effectiveness.

The analysis of the quantity theory of money is next explored, which in practice refers to the analysis of the monetary policy of banks. The current mainstream of economists believes that the currency exchange rate can be stabilized as long as there is confidence in the credit of the state so that the stability of the value of the monetary unit is realized. In other words, stabilization of the fixed exchange value with other currencies is considered first and foremost by monetary policy, whose main objective is external, rather than internal, convertibility. When the Bretton Woods system collapsed, meaning that the US abandoned its commitment to convertibility between the dollar and gold, the link between money and fixed metals was completely severed. Friedman once said, "The real-world monetary system is unprecedented. There is no longer a link between any kind of money and commodities". Thus, we now live in a world that does not need a monetary entity, which is backed by the credit of virtualized money. And the monetary policy of banks refers to the fractional reserve system mentioned in the previous chapter.

In traditional economic theory, Keynesians have been pushing up market demand through an excess supply of money. However, the reality is that increasing uncertainty in the money market has led to the foundations of the quantity theory of money being shaken. And the expectation that central banks regulate financial markets through the money multiplier, as perceived by monetarists, has been increasingly questioned. In the monetary theory of neoclassical economics, the rational allocation of resources can be achieved through the relative price mechanism of the equilibrium system of Walras. Under the premise that money is neutral, complex equations can be derived by calculating the relationship between money prices and monetary aggregates in order to obtain the corresponding aggregates. In short, the roots of the quantity theory of money are as follows.

Firstly, money is considered to be neutral. Therefore, the quantity of money can be calculated as a function of the quantity of money and the price level, which is the basis of all stable monetary theories.

Secondly, money is considered to be a durable capital asset, which can also be seen as an objective existence of value. In other words, the intrinsic value of money depends on the objective nature of money.

Thirdly, the speed of money circulation can be represented by a stable function. When rising inflation rates and financial innovation bring about huge changes, the nature of the market is one of uncertainty and instability in the supply of money. Money circulation expands without constraint, driven by leveraging. The quantity theory of money, which presupposes a stable demand for money and a determined quantity of money, is increasingly difficult to stand on.

Finally, topics related to price signals will be discussed. In a world dominated by finance capitalism, the allocation of resources, the structure of the economy, the development of industries and the distribution of income in each country are determined by three key price signals. These three price signals include interest rates, exchange rates and asset prices. The existence of these three signals not only enables the entire economic ecology to realize the value of virtualization, but also poses the risk of financial crises. These three signals are introduced into the network of the open economy, which leads to the following conclusions.

Firstly, financial markets are the most important markets in the entire economy in the era of financial capitalism. The global economic ecosystem is dominated by the financial markets represented by the money market, the bond market and the stock market. Therefore, our research also focuses on the inner logic of digital financial markets and, in particular, on the theory of innovation in digital finance. The core financial innovations that emerged as a result of the third industrial revolution refer to venture capital and the GEM market (which is represented by NASDAQ), private equity (PE) and equity incentives, the junk bond market and asset securitization. The financial revolution in the digital economy, on the other hand, is the digital currency, distributed ledger and Token economy ecosystem derived from blockchain technology.

Secondly, the idea of subjective value theory has been discussed before. All three price signals are also subjective variables, in the same way as the concepts of value, utility, wealth, costs and benefits. The subjective judgements and psychological expectations of the participants in the economic ecology are reflected by them. All these variables are relative variables and relative prices. The search for an equilibrium interest rate, an equilibrium exchange rate and an equilibrium asset price is therefore futile. The internal ecology of the financial system is therefore to be understood as a complex system operating under uncertain price signals.

Finally, from a digital economics perspective, the consideration of price signals in the financial ecosystem actually refers to the consideration of shareholder value or corporate value of the participants. In the real economic ecosystem, shareholder value is captured in the stock market. In the digital economy, on the other hand, shareholder value is reflected in the price of Tokens. In other words, a whole set of financial logic is built up by the traditional equity market with a market value culture or shareholder value culture. Digital finance, then, is likely to be a whole ecology built on the value of Tokens organized by corporate networks.

Token Economics and Price Theory

After the issue of trading networks and cryptocurrency in an open economy has been discussed, we return to the topic of the Token economy. We have been emphasising before how to establish a positive market economy in a blockchain trading network through a Token economy, and the role that Tokens assume in this closed network is that of a currency, especially in terms of price signals. Therefore, we need to derive the basic price logic of the Token economy by examining the theory of market prices. For the sake of space, we will focus here on the economic principles and ideas involved, rather than on the analysis of the specific mathematical expressions. As in the rest of the book, we will not only discuss the economic principles, but also refine them through other interdisciplinary theories.

The basic point we need to emphasise here is that only digital currencies issued by central banks have the properties of digital currencies. Blockchain technology is essentially a reliable technical trust mechanism for all parties to a transaction through mathematical algorithms, characterized by an algorithm that creates a ledger that can be shared and updated by all. Based on our previous analysis, we believe that neither "stable coins" nor other "digital currencies" issued by private institutions have the credit of a real fiat currency, but only digital currencies issued by the central bank are real currencies and have all the qualities that a local currency Only a digital currency issued by a central bank is a real currency and has all the qualities that a local currency should have. Only a digital currency issued by a central bank is a real currency and has all the qualities of a local currency. A privately issued currency cannot function as a fixed general equivalent because it is not fiat and compulsory, so it cannot function as a real circulating payment. Therefore, the digital currency we are discussing here refers to the digital currency issued by the central bank, while the Token is the general equivalent "with certain monetary properties".

Firstly, we discuss the relationship between monetary theory and the price level, which is the basic logic for establishing the monetary theory of the Token economy and the market price mechanism. Here it is time to discuss the research work related to the theory of the quantity of money in economics, from the British economist Hume who proposed this concept of thought, to the later Fisher who proposed the cash transaction theory, the British Cambridge School founder Marshall who proposed the cash balance theory and later Keynes and Friedman, all of whom analyzed the quantity of money. For the sake of space, we will briefly discuss the logic of the idea here, without getting into the quantitative calculations involved.

Early quantity theory of money was dominated by mercantilists and classical economists, including Bodin, Locke, Hume, Ricardo and Mill, who all published briefs on the quantity of money. Overall the early quantity theory of money this argued that there was a clear correlation between the quantity of money and the price of commodities and the value of money. For example, Locke developed the quantity theory of money in terms of the supply and demand for money, arguing that the value of money was high because the demand for money was greater than the supply. Montesquieu argued that silver coins are symbols of commodities and commodities are symbols of silver coins, and that they maintain a relative value relationship with each other, so that a change in either of them must result in a change in their value. Hume argued that the money in circulation in a country was merely a symbol of value used to calculate or represent commodities, and that the price of commodities was not determined by the absolute quantity of all commodities and money in a country, but by the quantity of commodities listed or to be listed and the quantity of money in circulation, and that an increase in the quantity of money, while the quantity of commodities remained the same, would increase the price of commodities in proportion. And the mastermind was Mill, who argued from a mechanistic point of view that the value of money depends on both the supply and demand for money and the quantity of goods in the market, of which the speed of money circulation has a clear impact on the value of money.

In the modern West, the quantity theory of money focused on the study of quantitative relationships in addition to the nature of the relationship between commodity prices and the value of money, represented by the economists Fisher, Marshall, Pigou, Keynes and Friedman. Let us take the example of Friedman's proposed money demand function, which uses a complex functional equation to show three basic logic. Firstly, the demand for money and its determinants can be analyzed using a stable continuity equation, i.e. a static equation that allows for the study of the quantity of money. Secondly, the demand for money is independent of the supply of money, i.e. the factors affecting the demand for money are completely independent of the supply of money, and price movements depend on the supply of money. Thirdly, there is a clear time lag between the growth of the quantity of money and the growth of nominal income, where the important influences include prices and output.

In short, both early and recent economists have argued that the core of the quantity theory of money is the relationship between the quantity of money and the price level, but so far no clear and accepted functional relationship between the two has been established, especially with the break-up of the Bretton Woods system and the development of the digital economy, and the need to find new functional relationships within it, especially the work on complex functions of the money supply, which is the objective of the analysis we will later carry out through the study of information theory and complex systems.

Then, we derive the price theory of the Token economy through the study of complex economic systems. By looking at the development of price theory and the study of complex economic systems, we sort out the basic ideas of price theory in the Token economy. Starting from the theory of complex economic systems, if we consider the economic ecology as a system, we can get a completely different perspective if we analyse it from the perspective of information theory. On the one hand, we already know the major flaw of traditional economics, which reduces the market to a homogeneous sum of quantities of a single commodity or "rational man", a logic that completely ignores the subjectivity of man and the heterogeneity of capital. On the other hand, we have already discussed the premise of exchange value and the market for transactions as a premise for analysis, so it is possible to study it through the logic of information theory, that is, as a complex system formed by the information of a large network of random transactions, in which price signals are formed. This logic is also in line with Adam Smith's metaphor of the "invisible hand", which sees the price system as the baton of the market. In the Token economy, such a logic would be more in line with the self-organized economic order of the blockchain network.

We are here to discuss one of the reasons for making Token prices one of the central studies of the entire digital economy ecology, and here it is necessary to introduce the central theory of financial capitalism: the efficient market hypothesis. On October 14, 2013, the University of Chicago added Eugene F. Farmer, Lars Henson, and Yale University Professor Robert Henson. Hansen and Yale University Professor Robert Schiller were invited to give a lecture on the subject. The Nobel Prize in Economics was awarded to Eugene F. Farmer, Lars Hansen and Robert Shiller for their empirical analysis of asset price volatility, with Professor Farmer as the pioneer of the Efficient Market Hypothesis. The efficient market hypothesis states that asset prices (the prices of stocks and bonds) perfectly reflect all information related to the price, value and future income stream of an asset, i.e. the price of an asset resulting from future purchases and sales perfectly covers all information related to the price, value and future income stream of an asset, so the study of the price of financial assets is essentially a study of market mechanisms. Therefore, the study of market prices in the Token economy is also the focus of the analysis of the entire Token economy ecology.

Finally, we come to the analysis of the application of information economics theory in blockchain transaction networks. Here we start with a discussion of the main theoretical traces and basic applications. Here is the main American economist and 1982 Nobel Prize winner in economics George. Stigler's work on information economics and 2001 Nobel Prize in Economics winner George W. Bush. Akerlof, Akerlof, and Joseph. Stiglitz's theory of information market research as the basis.

Stigler, a representative of the Chicago School in microscience and one of the founders of information economics, argued that the cost of information to consumers in obtaining the quality, price and timing of purchases of goods was so great that buyers neither could nor wanted to be adequately informed, resulting in the existence of different prices for the same good. Stigler argues that this is an inevitable, normal market and market phenomenon that does not require human intervention. Stigler's view updates the assumption in microeconomic market theory that there is only one price for a commodity. In the course of his research, Stigler also extended this analysis to the labour market, and these studies created a new field of research known as "The Economics of Information" [41]. The other two were both awarded the 2001 Nobel Prize in Economics, with Akerlof's contribution being to illustrate the market reality that the ability of sellers to sell low-quality goods to buyers exists because of asymmetries in the information available to each side of the market, and Spence's research revealing the theory of how people can make more money by having more information. Stiglitz, on the other hand, elaborates on the theory of how the less informed side of the market can adjust to the market. In short, their work in the area of "analysis of markets full of asymmetric information" is worth noting.

In the study of the Token economy, we are about to see the following three findings on the basic order of the Token economy.

Firstly, it can be understood through the mechanism of price diffusion in the market. Token's price signals are built on a free market order that allows for different price signals for homogeneous goods, as well as the existence of different types of price signals in multiple networks. Token mechanisms share some similarities with monetary mechanisms mainly in terms of price mechanisms, and indeed The Token economy is a complex system related to enhancements and incentives. Tokens can be used in niche economies as an enhancing and incentivising medium of exchange for value, allowing for greater scalability of niche economies through positive incentives for value circulation. How positive incentive behaviour is enhanced by Tokens for all parties involved is fundamental to the analysis of Tokens, and price mechanisms are an important means of understanding and quantifying their effects.

Secondly, it can be understood through the mechanism of information asymmetry. How to dissolve information asymmetry in a network is the mechanism that can be built into a blockchain network, and the Token is an expression of the price of an economic ecology under information symmetry. This part is therefore more in line with the basic assumptions of the efficient price theory we discussed, namely that the cost of information acquisition is zero as well as that market regulation is instantaneous. As Tokens are circulated in the blockchain network, they are secure, efficient and quick to circulate, requiring only a very short time and very few resources to circulate in the network. Compared to previous digital assets in the digital economy (e.g. points, vouchers, notes, etc.), they offer a higher level of value security and trustworthiness, which is the basis for our analysis of the circulation of Token value based on information theory. Moreover, as Tokens can be value-matched to multiple enhancements, the fluctuation of Tokens can accommodate multiple dynamic price mechanisms, and the price of each Token will gain almost instant certainty in different scenarios, which is more immediate and effective compared to the price signals of the market.

Finally, competition in the case of information symmetry lies in the differentiation of goods in terms of experience and service. Therefore, we can expect the market in a blockchain network to be an economic model with service and experience at its core. How to build a complex business ecosystem with an efficient experience economy through the reshaping of the economic order is a key concern for every blockchain network when designing Tokens. As the same economic system unfolds and expands around smart contracts, the properties of smart contracts can be fully reflected in the use of Tokens. We understand that the greatest value of smart contracts is that any individual, organization or institution can initiate entitlements based on their own resources and consensus, and that these entitlements correspond to Tokens that are verifiable, traceable and exchangeable, and as we have previously discussed the intrinsic link between smart contracts and contract economics, Tokens based on technical contracts allow for a more efficient and trustworthy complex network organization.

The above is our study of the Token economy and price theory, through a paradigm shift in the study of money and price theory, i.e. from a single neoclassical economics of price theory to a study in a complex economic system with information theory at its core, and the introduction of the efficient market hypothesis and the thinking of information economics. We can see that the key question of the Token economy is how to establish a market mechanism for correct price signals in a self-organized market order, which is most importantly concerned with the logic of the construction of price signals and monetary ecology, i.e. the establishment of a framework for digital finance that is fully compatible with efficient price market theory.

In summary, questions about the economic order in open networks are explored in this chapter. It consists of three main themes, namely Transaction Network and Capital Structure, Open Economy and Monetary Ecology, and Token Economics and Price Theory. The limitations of traditional monetary and price theory in economics are seen. Therefore, theories of information theory and complex networks are used to discuss this area of research. At the same time, some basic ideas of information economics are introduced. In digital economics, the basic premise of the discussion on the theory of the Token economy became the problem of establishing a monetary mechanism with full symmetry of information. And the establishment of price signals and efficient markets is the key to be considered. When this analytical idea is understood, the differences between the digital economy and other traditional economic ecologies, and the basic framework of how economic order is established in this ecology, can be understood.

References

1. Harrod R (1952) An essay in dynamic theory. In: Economic essays. Palgrave Macmillan, London, pp 254–277
2. Solow RM (1956) A contribution to the theory of economic growth. Quart J Econ 70:65–94
3. Romer PM (1986) Increasing returns and long run growth. J Polit Econ 94(5):1002–1037. (36 pages) Published By: The University of Chicago Press
4. Evans G, Honkapojha S (1998) Growth cycles. Am Econ Rev 88(3):495–515
5. Maddison A (2003) The world economy: a millennial perspective. Peking University Press, Beijing
6. Aristotle (2003) The Nicomachean ethics. Commercial Press, Beijing. Translator Liao Shenbai
7. Hume D (1751) An enquiry concerning the principles of morals, vol 1. A. Millar, London [28 June 2014] via Google Books
8. Economic doctrine and method: an historical sketch
9. EUR-Lex—32016R0679—EN—EUR-Lex. Eur-lex.europa.eu [2018-05-28]
10. Schumpeter J (2015) The theory of economic development. Huaxia Publishing House, Beijing. Translated by Guo Wujun and Lv Yang
11. Arthur B (2014) The nature of technology. Zhejiang People's Publishing House, Hangzhou
12. Rostow WW (1960) The five stages of growth—a summary. In: The stages of economic growth: a non-communist manifesto. Cambridge University Press, Cambridge, pp 4–16. Archived from the original on 2013-02-23
13. Keynes JM (1936) The general theory of employment, interest, and money. Macmillan, London
14. Keynes JM (1930) A treatise on money. Harcourt, San Diego. ISBN: 978-0-404-15000-6
15. Marshall A (1964) Commenting on Alfred Marshall's principles of economics, principles of economics. Commercial Press, Beijing. ISBN: 9781573921404
16. Smith A (1776) An inquiry into the nature and causes of the wealth of nations. W. Strahan and T. Cadell, London
17. Newton SI (1687) Philosophiae Naturalis Principia Mathematica. J. Societatis Regiae ac Typis J. Streater, London
18. Hayek FA (2000) The fatal conceit: the errors of socialism. China Social Sciences Press, Beijing, p 247
19. Hobbes T (2017) Leviathan. Commercial Press, Beijing. Translated by Li Sifu and Li Tingbi, Proofread by Yang Changyu. ISBN: 9787100123556
20. Cantillon R (1997) Essay on the nature of commerce in general. Commercial Press, Beijing. ISBN: 9787100023757
21. Kerzner I (2013) Competition and entrepreneurship. Zhejiang University Press, Hangzhou. ISBN: 9787308110921

Z. Liu, *Principles of Digital Economics*, Contributions to Economics,
https://doi.org/10.1007/978-981-16-9020-4

22. Petty W (2013) Treatise on taxes and contributions. Huaxia Publishing House, Beijing. Translated by Qiu Xia and Yuan Lei. ISBN: 9787508076515
23. Bohm-Bawerk EV (2010) Capital and interest. Macmillan, London. ISBN: 9787100071673
24. Kelly K (1994) Out of control: the new biology of machines, social systems, and the economic world. Addison Wesley, Boston. Perseus Books, 1995
25. Arthur WB (2014) Complexity and the economy. Zhejiang People's Publishing Press, Hangzhou. Translated by Jia Yongmin. ISBN: 9787213086458
26. Moore JF (1999) The decline of competition: leadership and strategies in the age of business ecosystems. Beijing Publishing House, Beijing. ISBN: 9787200033571
27. Menger C (2013) Principles of economics. Truth & Wisdom Press, Shanghai. Translated by Liu Jieao. ISBN: 9787543222779
28. Von Mises L (2015) Human action: a treatise on economics. Shanghai Academy of Social Sciences Press, Shanghai. ISBN: 9787552008494
29. Christensen CM (2014) The innovator's dilemma. CITIC Publishing Press, Beijing. ISBN: 9787508642802
30. Collins J, Porras J (2006) Built to last: successful habits of visionary companies. CITIC Publishing Press, Beijing. ISBN: 9787800734762, x-9
31. Coase RH (1937) The nature of the firm. Economica 4(16):386–405
32. von Mises L (2018) The theory of money and credit. Commercial Press, Beijing. Translated by Fan Linzhou. ISBN: 9787100166706
33. Rothbard MN (2020) Man, economy and state. Zhejiang University Press, Hangzhou. ISBN: 9787308200486
34. North DC, Wallis JJ, Wingstedt BR (2017) Violence and social order. Truth & Wisdom Press, Shanghai. Translated by Hang Xing, Wang Liang. ISBN: 9787543218642
35. North DC (2014) In: Xin C (ed) Institutions, institutional change and economic performance. Truth & Wisdom Press, Shanghai. ISBN: 9787543223547
36. Rothbard M (2003) America's great depression. Shanghai People's Publishing Press, Shanghai. Translated by Xie Huayu. ISBN: 9787208047396
37. Lachmann L (2015) Capital and its structure. Shanghai University of Finance and Economics Press, Shanghai. ISBN: 9787564222697
38. Songzuo X (2014) The new capitalism. CITIC Publishing Press, Beijing. ISBN: 9787508649085
39. Yifu L (2012) New structural economics: a framework for rethinking development and policy. Peking University Press, Beijing. ISBN: 9787301174975
40. Burt R (2008) Structural hole: the social structure of competition. Truth & Wisdom Press, Shanghai. Translated by Ren Min/Li Lu/Lin Hong. ISBN: 9787543214309
41. Stigler GJ (1961) The economics of information. J Polit Econ 69(3):213–225

The manufacturer's authorised representative in the EU is Springer
Nature Customer Service Centre GmbH, Europaplatz 3, 69115 Heidelberg,
Germany. If you have any concerns regarding our products, please
contact ProductSafety@springernature.com

Printed and bound by CPI Group (UK) Ltd, Croydon, CR0 4YY
30/09/2026
02225264-0019